IRRIGATION ENGINEERING

IRRIGATION ENGINEERING

Editor

R.N. Reddy

2025

Gene-Tech Books

ISBN 978-81-89729-98-1

Published by : **GENE-TECH BOOKS**
4762-63/23, Ansari Road, Darya Ganj,
NEW DELHI - 110 002
Phone: 011 43003222
e-mail: genetechbooks@yahoo.co.in

PRINTED IN INDIA

Preface

Irrigation involves the artificial application of water to soil, usually for assisting the growth of crops in dry areas or where there is a shortage of rainfall. As the process of agriculture becomes increasingly mechanised, the application of scientific methods and technology to the procedure of irrigation comes as no surprise. With agricultural yields dwindling and demand for food increasing, the pressure on agriculture is immense and no stone is to be left unturned in meeting demands and expectations. Effective irrigation, enhanced pertinent technologies, thereby provides a most lucrative source for making agriculture profitable and worthwhile.

This text has been written as a manual guiding agricultural engineers on the principles and concepts which define irrigation engineering. Elaborating upon the tools, technologies and techniques which are essential to the field, the text takes a look at the types of irrigation, including surface, localised and sprinkler systems, and how they are changing agriculture itself by making it more scientific. In addition to it the challenges faced by irrigation systems including those of entrophication, water pollution, depletion of underground aquifers, etc., are also discussed. Current trends and development have been analysed as well as the future prospects and opportunities.

R.N. Reddy

Contents

	Preface	*v*
1.	Importance of Irrigation Engineering	1
2.	Estimation of Irrigation Demand	35
3.	Irrigation Schemes and Methods	47
4.	Modern Canal Operation Techniques	71
5.	Design of On-demand Irrigation Systems	111
6.	Mapping Water Networks	125
7.	Drainage Pipes and Accessories	143
8.	Drainage and Disposal Systems	179
9.	Surface and Subsurface Drainage	201
10.	Installation and Maintenance of Drainage Materials	239
11.	Evaluation of Irrigation Project	255
	Bibliography	*269*
	Index	*271*

1

Importance of Irrigation Engineering

Irrigation involves artificially providing crops with water. This technique is used in farming to enable plants to grow when there is not enough rain, particularly in arid areas. It is also used in less arid regions to provide plants with the water they need when seed setting. About 66% of the world's water catchment is used in farming, which continues to make increasing use of irrigation. But in most irrigation systems 50 to 60% of the water used does not benefit the plants. It is therefore necessary to set up more carefully designed irrigation schemes that ensure optimum agricultural production while preserving this resource.

The choice of irrigation method depends on the type of crop and economic context. For small areas a network of open channels providing water by gravity remains the least costly and simplest solution. More sophisticated modern techniques can reduce water consumption: drip irrigation delivers the water at the plants' roots and sprinkler irrigation sprays it over the crop.

For the survival of the country, there is an urgent need to implement and plan irrigation strategies for now, and in future, as the population continues to grow. But that should not be at the cost of degradation of the present available resources of land and water, which means the natural resources that we have, should more or less remain the same after 50 or 100 years and beyond. In many regions of India, there has been alarming withdrawal of ground water for meeting demands of irrigation and drinking water demand than that which can be naturally recharged. This has led to rise of further problems like arsenic and fluoride contamination. Since ground water recharge by natural means takes a long time, perhaps years and even decades, there is little hope of regaining the depleted table near future.

The total geographical area of land in India is about 329 million hectare (M-ha), which is 2.45% of the global land area. The total arable land, according to an estimate made by the Food and Agricultural Organisation (FAO), made available through the web-site Aquastat, is about 165.3 M-ha which is about 50.2 % of the total geographical area. India possesses 4% of the total average annual run off in the rivers of the world. The per capita water availability of natural run off is at least 1100 cubic meters/yr.

The utilisable surface water potential of India has been estimated to be 1869 cubic kms but the amount of water that can actually be to put to beneficial use is much less due to severe limitations imposed by physiographic, topographic, interstate issues and the present technology to harness water resources economically. The recent estimates made by the Central Water Commission indicate that the water resources utilisable through surface structures is about 690 cubic kms only (about 36% of the total ground water is another important source of water).

Ground water is another important source of water. The quantum of water that can be extracted economically from the ground water aquifers every year is generally known as ground water potential. The preliminary estimates made by the Central Ground Water Board indicate that the utilisable ground water is about 432 cubic km.thus the total utilisable water resource is estimated to 1122 cubic km. It must be remembered that this amount of water is unequally spread over the entire length and breadth of the country. Of the total 329 M-ha of land, it is estimated that only 266 M-ha possess potential for production. Of this, 143 M-ha is agricultural land. It is estimated that 85 M-ha of land suffers from varying degrees of soil degradation. Of the remaining 123 M-ha, 40 M-ha is completely unproductive. The balance 83 M-ha is classified as forest land, of which over half is denuded to various degrees.

It is alarming to note that the percapita availability of land is half of what it used to be same 35 years ago. This would further reduce as our country's population continuous to grow. At present 141 M-ha of land is being used for cultivation purposes. Between 1970-71 and 1987-88 the average net sown area has been 140.4 M-ha. The need for production of food, fodder, fibre, fuel in the crop growing areas have to compete with the growing space require for urbanisation. The factors of land degradation, like water logging, salinity, alkalinity and erosion of soils on account of inadequate planning and inefficient management of water resources projects will severely constrain the growth of net sown area in the future.

Benefits of Irrigation

With the introduction of irrigation, there have been many advantages, as compared to the total dependence on rainfall. These may be enumerated as under:

— *Increase in crop yield*: the production of almost all types of crops can be increased by providing the right amount of later at the right time, depending on its shape of growth. Such a controlled supply of water is possible only through irrigation.

— *Protection from famine*: the availability of irrigation facilities in any region ensures protection against failure of crops or famine due to drought. In regions without irrigation, farmers have to depend only on rains for growing crops and since the rains may not provide enough rainfall required for crop growing every year, the farmers are always faced with a risk.

— *Cultivation of superior crops*: with assured supply of water for irrigation, farmers may think of cultivating superior variety of crops or even other crops which yield high return. Production of these crops in rain-fed areas is not possible because even with the slight unavailability of timely water, these crops would die and all the money invested would be wasted.

— *Elimination of mixed cropping*: in rain-fed areas, farmers have a tendency to cultivate more than one type of crop in the same field such that even if one dies without the required amount of water, at least he would get the yield of the other. However, this reduces the overall production of the field. With assured water by irrigation, the farmer would go for only a single variety of crop in one field at anytime, which would increase the yield.

— *Economic development*: with assured irrigation, the farmers get higher returns by way of crop production throughout the year, the government in turn, benefits from the tax collected from the farmers in base of the irrigation facilities extended.

— *Hydro power generation*: usually, in canal system of irrigation, there are drops or differences in elevation of canal bed level at certain places. Although the drop may not be very high, this difference in elevation can be used successfully to generate electricity. Such small hydro electric generation projects, using bulb-turbines have been established in many canals, like Ganga canal, Sarada canal, Yamuna canal etc.

— *Domestic and industrial water supply*: some water from the irrigation canals may be utilised for domestic and industrial water supply for nearby areas. Compared to the irrigation water need, the water requirement for domestic and industrial uses is rather small and does not affect the total flow much.

Types of Irrigation Schemes

Irrigation projects in India are classified into three categories -major medium & minor according to the area cultivated the classification criteria is as follows:-

- — *Major irrigation projects*: projects which have a culturable command area (CCA) of more than 10,000 ha but more than 2,000 ha utilise mostly surface water resources.
- — *Medium irrigation projects*: projects which have CCA less than 10,000 ha. But more than 2,000 ha utilises mostly surface water resources.
- — *Minor irrigation projects*: projects with CCA less than or equal to 2,000 ha. utilises both ground water and local surface water resources. Ground water development is primarily done through individual and cooperative effort of farmers with the help of institutional finance and their own savings. Surface water minor irrigation schemes are generally funded from the public sector only.

The ultimate irrigation potential from minor irrigation schemes have been assessed as 75.84 million ha of which partly would be ground water based (58.46 million ha) and covers about two thirds. By the end of the ninth plan, the total potential created by minor irrigation was 60.41 million ha. The ultimate irrigation potential of the country from major and medium irrigation projects has been assessed as about 64 M-ha. By the end of the ninth plan period, the total potential created from major and medium projects was about 35 M-ha.

Role of Irrigation Projects

While formulating strategies for irrigation development the water resources planner should realise the benefits of each type of project based on the local conditions. For example, it may not always be possible to benefit remote areas using major/medium projects.

At these places minor irrigation schemes would be most suitable. Further, land holding may be divided in such a way that minor irrigation becomes inevitable. However, major and medium projects wherever possible is to be constructed to reduce the overall cost of development of irrigation potential. According to the third minor irrigation census carried out in

2000-01, there are about 5.56 lakh tanks in the country, with the most occurring in the following states

— West Bengal: 21.2 percent of all the tanks in the country
— Andhra Pradesh: 13.6
— Maharashtra: 12.5
— Chhattisgarh: 7.7
— Madhya Pradesh: 7.2
— Tamilnadu: 7.0
— Karnataka: 5.0

Due to non use of these 15 percent tanks nearly 1 M-ha of Irrigation potential is lost. Another, around 2 M-ha of potential is lost due to under utilisation of tanks in use. Loss of potential due to non use is more pronounced in Meghalaya, Rajasthan and Arunachal Pradesh (above 30%), whereas loss of potential due to under utilisation is more than 50 percent in case of Gujarat, Nagaland, Rajasthan, A&N Island and Dadar and Nagar Haveli.

It also appears that the maintenance of the tanks has been neglected in many parts of the country and their capacity has been reduced due to siltation. It has been estimated that about 1.7 M-ha of net area has been lost under tank irrigation due to drying up of tanks and encroachment of foreshore area. Some advantages of minor irrigation should also be kept in mind. These are: small investments, simpler components, labour intensive, quick maturing and most importantly it is farmer friendly.

On the other hand, it is seen that of the assessed 64 M-ha of irrigation potential that may be created through major and medium projects, only about 35 M-ha have so far been created. Hence a lot of scope for development in this sector is remaining. These may be realised through comprehensive schemes including storage, diversion and distribution structures. Some of these schemes could even be multipurpose thus serving other aspects like flood control and hydro power.

National Water Policy

Our country had adapted a national water policy in the year 1987 which was revised in 2002. The policy document lays down the fact that planning and development of water resources should be governed by the national perspective. Here quote the aspects related to irrigation from the policy.

- Irrigation planning either in an individual project or in a watershed as a whole should take into account the irrigability of land, cost-effective irrigation options possible from all available sources of water and appropriate irrigation techniques for optimising water use efficiency. Irrigation intensity should be such as to extend the benefits of irrigation to as large a number of farm families as possible, keeping in view the need to maximize production.
- There should be a close integration of water use and land use policies.
- Water allocation in an irrigation system should be done with due regard to equity and social justice. Disparities in the availability of water between head-reach and tail end farms and between large and small farms should be obviated by adoption of a rotational water distribution system and supply on a volumetric basis subject to certain ceilings and rational pricing.
- Concerted efforts should be made to ensure that the irrigation potential created is fully utilised. For this purpose, the command area development approach should be adopted in all irrigation projects.
- Irrigation being the largest consumer of fresh water, the aim should be to get optimal productivity per unit of water. Scientific management farm practices and sprinkler and drip system of irrigation should be adopted wherever feasible.
- Reclamation of water-logged/saline affected land by scientific and cost effective methods should form a part of command area development programme.

Command Area Development Programme (CADP)

This scheme, sponsored by the central government was launched in 1974-75 with the objective of bridging the gap between irrigation potential created and that utilised for ensuring efficient utilisation of created irrigation potential and increasing the agricultural productivity from irrigated lands on a sustainable basis. The programme envisages integrating various activities relating to irrigated agriculture through a multi-disciplinary team under an area development authority in a coordinated manner. The existing components of the CADP are as follows:-

- — On farm development works, that is, development of field channels and field drains within the command of each outlet, land levelling on an outlet command basis; reclamation of water logged areas; enforcement of a proper system of rotational water supply (like the warabandi) and fair distribution of water to individual fields; realignment of field boundaries, wherever necessary (where possible, consolidation of holding are also combined) supply of all inputs and service including credit; strengthening of extension services; and encouraging farmers for participatory irrigation management.
- — Selection and introduction of suitable cropping patterns.
- — Development of ground water to supplement surface irrigation (conjunctive use under minor irrigation sector)
- — Development and maintenance of the main and intermediate drainage system.
- — Modernisation, maintenance and efficient operation of the irrigation system up to the outlet of one cusec (1ft3/ sec) capacity.

For an overall appreciation of an entire irrigation project it is essential that the objectives of the CAD be kept in mind by the water resources engineer.

Participatory Irrigation Management (PIM)

Any irrigation project cannot be successful unless it is linked to the stakeholders, that is, the farmers themselves. In fact, people's participation in renovation and maintenance of field channels was the established practice during the pre independence days. However, the bureaucracy encroached on this function in the post independence period and a realisation has dawned that without the participation of farmers, the full potential of an irrigation scheme may not be realised. Though a water resources engineer is not directly involved in such a scheme, it is nevertheless wise to appreciate the motive behind PIM and keep that in mind while designing an irrigation system. The national water policy stresses the participatory approach in water resources management. It has been recognised that participation of the beneficiaries would help greatly for the optimal upkeep of irrigation system and utilisation of irrigation water.

The participation of farmers in the management of irrigation would give responsibility for operation and maintenance and collection of water rates from the areas under the jurisdiction of the water user's association of concerned hydraulic level. Under the command area development programme (CADP), presently a provision exists for a one-time functional grant to farmer's associations at the rate of Rs. 500 per hectare of which Rs. 225 per hectare is provided by the central government and state government each, and Rs. 50 per hectare is contributed by the farmer's associations. It may be mentioned that so far, that is by year 2004, the state governments of Andhra Pradesh, Goa, Karnataka, Tamilnadu, Rajasthan and Madhya Pradesh have enacted legislations for the establishment of the water user's associations.

The sustainability and success of PIM depends on mutual accountability between the water user's association and the irrigation department of the concerned state, attitudinal change in the bureaucracy, autonomy for the water user's associations, multifunctional nature of the water user's association and the choice of appropriate model for PIM with appropriate legal and institutional framework. If the farmers have to take over and manage the system, then the system must be rectified by the irrigation department to a minimum standard to carry the design discharge before it is handed over to the water user's association. The success of the PIM is also linked to the introduction of rotational water supply and water charges with rationalised establishment costs. Unlined field channels need to be manually constructed in a 'V' shape which is considered stable and efficient for carrying water.

Irrigation Water Management

Of the two resources -land and water, management of the former is largely in the domain of agricultural engineers. Management of water, on the other hand, is mostly the purview of the water resources engineer who has to decide the following:

— How much water is available at a point of a surface water source, like a river

— How much ground water is available for utilisation in irrigation system without adversely lowering the ground water table?

— For the surface water source, is there a need for construction of a reservoir for storing the monsoon runoff to be used in the lean seasons?

— What kind of diversion system can be constructed across the river for diverting part of the river flow into a system of canal network for irrigating the fields?

— How efficient a canal network system may be designed such that there is minimum loss of water and maximum agricultural production?

— How can excess water of an irrigated agricultural fields be removed which would otherwise cause water logging of the fields?

In order to find proper solution to these and other related issues, the water resources engineer should be aware of a number of components essential for proper management of water in an irrigation system. These are:-

Watershed Development

Since the water flowing into a river is from a watershed, it is essential that the movement of water over ground has to be delayed. This would ensure that the rain water falling within the catchment recharges the ground water, which in turn replenishes the water inflow to the reservoir even during the lean season. Small check dams constructed across small streams within the catchment can help to delay the surface water movement in the watershed and recharge the ground water. Measures for the water shed development also includes a forestation within the catchment area which is helpful in preventing the valuable topsoil from getting eroded and thus is helpful also in preventing siltation of reservoirs. Other soil conservation methods like regrassing and grass land cultivation process, galley plugging, nullah bunding, contour bunding etc. also come under watershed development.

Water Management

Surface water reservoirs are common in irrigation systems and these are designed and operated to cater to crop water requirement throughout the year. It is essential, therefore, to check loss of water in reservoir due to

— Evaporation from the water surface

— Seepage from the base

— Reduction of storage capacity due to sedimentation

Water Management in Conveyance System

In India the water loss due to evaporation, seepage and mismanagement in the conveyance channels (for canals and its distributaries) is exceptionally high-nearly 60%. Some countries like Israel have reduced this loss tremendously by taking several measures like lining of water courses, lining not only reduces seepage, but also minimizes weed infestation and reduces overall maintenance cost though the initial cost of providing lining could be high depending on the material selected.

On Farm Water Management

Though this work essentially is tackled by agricultural engineers, the water resources engineers must also be aware of the problem so that a proper integrated management strategy for conveyance-delivery-distribution of irrigation water is achieved. It has been observed from field that the water delivered from the canal system to the agricultural fields are utilised better in the head reaches and by the time it reaches the tail end, its quantity reduces. Often, there are land holding belonging to different farmers along the route of the water course and there is a tendency of excess withdrawal by the farmers at the upper reaches.

In order to tackle this kind of mismanagement a proper water distribution roster has to be implemented with the help of farmers' cooperatives or water user's associations. At times farmers are of the opinion that more the water applied more would be the crop production which is generally not true beyond a certain optimum water application rate. Education of farmers in this regard would also ensure better on-farm water management.

Choice of Irrigation Method

Though irrigation has been practiced in India from about the time of the Harappa civilisation, scientific irrigation based on time variant crop water need within the constraints of water and land availability is rather recent. It is important to select the right kind of irrigation method to suit the particular crop and soil. For example, following is a short list of available methods corresponding to the kind of crop.

Method of irrigation	Suitable for crops
Border strip method	Wheat, Leafy vegetables, Fodders
Furrow method	Cotton, Sugarcane, Potatoes
Basin method	Orchard trees

Other methods like sprinkler and drip irrigation systems are adapted where water is scarce and priority for its conservation is more than the consideration for cost. Although most advanced countries are adopting these measures, they have not picked up as much in India mainly due to financial constraints. However, as time passes and land and water resources get scarce, it would be essential to adopt these practices in India, too.

Choice of Cropping Pattern

Scientific choice of cropping pattern should be evolved on the basis of water availability, soil type, and regional agro-climate conditions. Crop varieties which give equivalent yield with less water requirement and take less time to mature should be popularised. Scientific contribution in the form of double or multiple cropping can be achieved if the sowing of crops such as paddy, groundnut, arhar etc. is advanced, if necessary, by raising the nurseries with the help of groundwater. Selection of crops planting sequence per unit weight of water applied.

Scheduling of Irrigation Water

Traditional farmers engaged in crop production are aware of some kind of scheduling of water to the crops, but their knowledge is based mostly on intuition and traditional wisdom rather than on any scientific basis. Modern scientific study on crop growth has shown that a correlation can be established between the climatic parameters, crop water requirement and the moisture stored in the soil especially in the root zone. It has now been established by scientific research that the application of irrigation should be such that the available water in the soil above the permanent wilting point is fully utilised by the crop before requiring application of water to replenish the depleted moisture in the soil.

Since any canal would be delivering water at the same time to different fields growing different crops, the demand of the various fields have to be calculated at any point of time or a certain period of time (days, weeks), and the water distributed accordingly through the canal network.

Development of Land Drainage

Due to improper application of water and inadequate facilities for drainage of excess water from irrigated lands, large tracks of land near irrigated areas have been affected with water logging and excess salt concentration in soils. Adequate drainage measures like surface and subsurface drainage systems, vertical drainage, bio-drainage etc. should be developed as an integral part of the irrigation system.

Command area Development

We have already seen that the government has initiated the command area development programme (CADP) which would ensure efficient water utilisation and integrated area developments in the irrigation command.

Canal Automation

At present, the water entering the canal network through the headworks as well as the water getting distributed into the various branches and finally reaching the fields through the outlets are controlled manually. However, if these operations are carried out through automated electromechanical systems which can communicate to a central computer, then the whole process can be made more efficient. This would also help to save water and provide optimal utilisation of the availability water.

Classification of Irrigation Waterchemes

The classification of the irrigation systems can also be based on the way the water is applied to the agricultural land as:

— *Flow irrigation system*: where the irrigation water is conveyed by growing to the irrigated land. This may again be classified into the following. Direct irrigation: Where the irrigation water is obtained directly from the river, without any intermediate storage. This type of irrigation is possible by constructing a weir or a barrage across a river to raise the level of the river water and thus divert some portion of the river flow through an adjacent canal, where the flow takes place by gravity.

Reservoir/tank/storage irrigation: The irrigation water is obtained from a river, where storage has been created by construction an obstruction across the river, like a dam. This ensures that even when there is no inflow into the river from the catchment, there is enough stored water which can continue to irrigate fields through a system of canals.

— *Lift irrigation system*: Where the irrigation water is available at a level lower than that of the land to be irrigated and hence the water is lifted up by pumps or by other mechanical devices for lifting water and conveyed to the agricultural land through channels flowing under gravity.

Classification of irrigation systems may also be made on the basis of duration of the applied water, like:

— *Inundation/flooding type irrigation system*: In which large quantities of water flowing in a river during floods is allowed to inundate the land to be cultivated, thereby saturating the soil. The excess water is then drained off and the land is used for cultivation. This type of irrigation uses the flood water of rivers and therefore is limited to a certain time of the year. It is also common in the areas near river deltas, where the slope of the river and land is small. Unfortunately, many of the rivers, which were earlier used for flood inundation along their banks, have been embanked in the past century and thus this practice of irrigation has dwindled.

— *Perennial irrigation system*: In which irrigation water is supplied according to the crop water requirement at regular intervals, throughout the life cycle of the crop. The water for such irrigation may be obtained from rivers or from walls. Hence, the source may be either surface or ground water and the application of water may be by flow or lift irrigation systems.

Pricing of Water

This is more of a management issue than a technical one. After all, the water being supplied for irrigation has a production cost which has to be met from either the beneficiary or as subsidy from the government. Since water is a state subject, every state independently fixes the rates of water that it charges from the beneficiaries, the remaining being provided from state exchequer. There are wide variations in water rate structures across states and the rate per unit volume of water consumed varies greatly with the crop being produced.

The rates charged in some states for irrigation vary even for different projects depending on the mode of irrigation. The rates, at present, also vary widely for the same crop in the same state depending on irrigation season, type of system, etc. As such, right now, there is no uniform set principles in fixing the water rates and a wide variety of principles for pricing are followed for the different states, such as:

- — recovery cost of water
- — capacity of irrigators to pay based on gross earning or net benefit of irrigation
- — water requirement of crops
- — sources of water supply and assurance
- — classification of land linked with land revenue system

In some states water cess, betterment levy, etc. are also charged. Hence, there is an urgent need of the water resources planners to work out a uniform principle of pricing irrigation water throughout the country, taking into account all the variables and constraints.

Major and Medium Irrigation Project Scheme in India

The central design organisation of each state desiring to set up an irrigation project shall have to prepare a detailed project report of the proposed irrigation scheme based on the document "Guidelines for Submission, Appraisal and Clearance of Irrigation and Multipurpose Projects" brought out by the Central Water Commission. This report has to be sent to the project appraisal organisation of the Central Water Commission for the clearance with a note certifying the following:

— All necessary surveys and investigations for planning of the project and establishing its economic feasibility have been carried out according to the guidelines mentioned above.

— 10% or 5000 ha (whichever is minimum) of the command area of the proposed project has been investigated in details in three patches of land representing terrain conditions in the command for estimation of the conveyance system up to the last farm outlets.

— 10% of the canal structures have been investigated in full detail.

— Detailed hydrological, geological, construction material investigations have been carried out for all major structures, that is, dams, weirs (or barrages, as the case may be), main canal, branch canal up to distributaries carrying a discharge of 10 cumecs.

— Soil survey of the command area has been carried out as per IS 5510-1969.

— Necessary designs for the various components of the project have been done in accordance with the guidelines and relevant Indian standards.

— Necessary studies for utilisation of ground water have been done with special regard to the problem of water logging and suitable provisions have been made for conjunctive use of ground water and drainage arrangements.

— The cropping pattern has been adopted in consultation with the state agriculture department and is based on soil surveys of the command keeping in view the national policy in respect of encouraging crops for producing oil seeds and pulses.

— The cost estimates and economic evaluations were carried out as per guidelines issued by the Central Water Commission.

It may be noted that similar report has to be made even for multipurpose projects having irrigation as a component. Apart from the above techno-economic studies carried out by the state central design organisation, the project report should be examined by the state-level project appraisal/technical advisory committee comprising representatives of irrigation, agriculture, fisheries, forests, soil conservation, ground water, revenue and finance departments and state environmental management authority. It may be noted that a project of the magnitude of a major or medium irrigation scheme has wide impacts.

Hence, the techno-economic feasibility report should also be supplemented with "Environmental Impact Assessment Report" and "Relief and Rehabilitation Plan". The latter touches the issue of the plans for appropriate compensation to the affected persons due to the construction of the projects. The project proposal submitted to the Central Water Commission shall be circulated amongst the members of the advisory committee of the ministry of water resources for scrutiny. Once the project is found acceptable it shall be recommended for investment clearance to the planning commission and inclusion in the five year plan/annual plan.

Plan Developmental Schemes in India

Since the commencement of the first five year plan in 1951, the developmental schemes of India have been planned in a systematic programme. Under this scheme, each state of the union of India has seen development in the field of irrigation development.

Andhra Pradesh

At the beginning of plan development there were about 56

major and medium projects in addition to several minor projects in the state. Irrigation potential of about 2.4 M-ha was created through these projects which was about one-fourth of the ultimate irrigation potential of the state. About one hundred major and medium projects were undertaken during plan development. These along with a large number of minor irrigation projects created an additional potential of about 4M-ha thus raising the irrigation potential created to about two-thirds of ultimate potential of 9.2 M-ha. This has resulted in bringing a net area of 4.3 M-ha under irrigation out of 11 M-ha of net sown area in the state. The important projects taken up in the last 50 years include large projects like Nagarjuna Sagar, Sriram Sagar and Srisailam in addition to projects like Tungabhadra high level canal, Somasila etc.

Arunachal Pradesh

Situated in the hilly region of the North east, irrigation potential of the state is as low as 0.16 M-ha. There was hardly any irrigation projects have been undertaken which have created an irrigation potential of 82,000 ha, which accounts for about half of the ultimate potential. A net area of about 36,000 ha out of the net sown area of 149000 ha has been brought under irrigation by these projects.

Assam

Before the commencement of plan development, irrigation development in the state was limited to minor projects utilising surface and ground water which provided an irrigation potential of 0.23 M-ha under plan development about 20 major and medium irrigation projects and a large number of minor irrigation projects were undertaken resulting in creation of an irrigation potential of 0.85 M-ha which is about one-third of the ultimate irrigation potential 2.67 M-ha for the state. A net area of 0.57 M-ha has been brought under irrigation through these projects. Important taken up during the period include Bodikerai, Dhansiri, Koilong etc.

Bihar

Irrigation development in the pre-plan period included four major and medium projects in addition to several minor irrigation projects. These projects accounted for an irrigation potential of 1.4 M-ha. About 105 major and medium irrigation projects in addition to a large number of minor irrigation projects were undertaken during plan development, resulting in creation of additional irrigational potential of about 7V. Two-third of the ultimate irrigation potential of 12.4 M-ha for the state has thus been created so far. An area of 3.3 M-ha out of net sown area of 7.7 M-ha in the state has been brought under irrigation from these projects. Important projects undertaken during the plan period include Gandak, Kosi barrage and Eastern canal, Rajpur canal, Sone high level canal, Subernarekha, North Koel reservoir etc.

Goa

Pre-plan irrigation development in Goa was limited to construction of a few minor projects. During plan development, four major and medium irrigation projects and a number of minor irrigation projects were undertaken. These projects have created an irrigation potential of 35000 ha, which accounts for about 43 percent of the ultimate irrigation potential of 82000 ha for the state. A net area of 23000 ha of the sown area of 131000 ha available to the state has been brought under irrigation. The important projects undertaken during the period include Salauli and Anjunem.

Gujarat

Two major projects and a number of minor irrigation projects were undertaken in the state during the pre-plan period which created an irrigation potential of 0.46 M-ha. About 130 major and medium irrigation projects were undertaken during plan development and these projects have created an irrigation potential of 3.4 M-ha Thus 74 percent of the ultimate irrigation potential of 4.75 M-ha for the state has been created. An area of 2.64 M-ha out of net sown area of 9.3 M-ha has been brought under irrigation under the above projects. Important projects taken up in the state include Ukai, Kakarpur, Mahi etc. The giant project of Sardar Sarovar is one of the projects presently ongoing in the state.

Harayana

The major project of Western Yamuna canal was constructed in the state during preplan period. Irrigation potential created in the pre-plan period was about 0.72 M-ha. About 10 major projects were taken up during the plan period in addition to a number of tubewells and other minor irrigation projects. The total irrigation potential created by projects so far undertaken amounts to 3.7 M-ha which accounts for about 80 percent of the ultimate irrigation potential for the state. Out of the net sown area of 3.6 M-ha, an area of 2.6 M-ha has been brought under irrigation. Important projects undertaken during the plan period include interstate projects like Bhakra-Nangal, Sutlej-Yamuna link canal etc.

Himachal Pradesh

Irrigation development in the hilly state of Himachal Pradesh was restricted to minor projects in the preplan period. During the plan development, 5 major and medium projects and a number of minor projects were undertaken in the state. Sofar, these projects have created a total irrigation potential of 0.16 M-ha and have brought an area of 99,000 ha under irrigation out of the net sown area of 5,83,000 ha in the state. Important projects undertaken in the State include Balh valley, Shahanahar etc.

Jammu & Kashmir

Pre-plan irrigation development in the state included seven major and medium projects in addition to minor irrigation schemes. These projects accounted for creation of an irrigation potential of 0.33 M-ha. During the plan development, about 20 major and medium irrigation projects and a number of minor projects were undertaken in the state. With the addition of about 0.22 M-ha from these projects the irrigation potential, so far, created has risen to 0.55 M-ha which is about 70 percent of the ultimate irrigation potential of 0.8 M-ha Net area brought under irrigation is 0.3 M-ha Important projects undertaken during plan period include Kathua Canal, Ravi-Tawi Lift, Karwal etc.

Karnataka

Pre-plan irrigation development in the state included 11 major and medium projects, in addition to a large number of minor projects. These projects created an irrigation potential of 0.75 M-ha. During plan period about 54 major and medium irrigation projects and a number of minor irrigation projects were undertaken. These projects have raised the irrigation potential created in the state to 0.32 M-ha which is about 70 percent of the ultimate irrigation potential of the state. Net area brought under irrigation 2.1 M-ha Important projects undertaken during the plan period include Ghataprabha, Malaprabha, Tungabhadra, Upper Krishna stage-I, Kabini, Harangi, Hemavati etc.

Kerala

Irrigation development in the state in the pre-plan period was limited to minor irrigation which had created a potential of 2,25,000 ha. During plan development, about 22 major and medium irrigation projects were undertaken which have raised the irrigation potential created to about 1.2 M-ha thus achieving 70 percent of the ultimate potential of 2.1 M-ha for the state. Net area brought under irrigation is 0.33 M-ha. Important projects taken up during plan development include Malampuzha, Chalakudi, Periyar Valley, Kallada etc.

Madhya Pradesh

A little over 1 M-ha of irrigation potential was created in the state in the pre-plan period through the construction of about 20 major and medium and a number of minor projects. During the period of plan development about 160 major and medium projects were undertaken along with minor irrigation schemes. With this, the irrigation potential created has gone upto over 5 M-ha which is about half of the ultimate potential of 10.2 M-ha. About 4.8 M-ha of land has been brought under irrigation through these projects. Important projects undertaken during the period include Chambal, Barna, Tawa, mahanadi Reservoir, Hasdeo-Bango, Bargi, Upper Wainganga etc.

Maharashtra

Irrigation development in pre-plan period in Maharashtra was also over 1 M-ha achieved through the construction of about 21 major and medium irrigation projects along with a number of minor irrigation projects. During plan development, over 250 major and medium projects and a large number of minor projects were added which raised the irrigation potential created to about 4.9 M-ha, which is about two-third of the ultimate potential of 7.3 M-ha Out of the net sown area of 18 M-ha available to the state, 2.5 Mha has been brought under irrigation. The important projects undertaken during the period of the plan development included Jayakwadi, Pench, Bhima, Mula, Purna, Khadakwasla, Upper Penganga etc.

Manipur

An irrigation potential of about 5,000 M-ha was created in the state during pre-plan period through minor irrigation projects. During plan development, six major and medium projects and a number of minor irrigation projects were added which has raised the irrigation potential to 0.14 M-ha which is about 60 percent of the ultimate potential of 0.24 M-ha A net area of 65,000 ha has been brought under irrigation through these projects. One of the important irrigation projects taken up is the Loktak lift.

Meghalaya

Minor irrigation projects are source of irrigation in this hilly state. A potential of 7,000 ha developed in the pre-plan period was enhanced to about 53,000 ha by taking up more minor irrigation projects. The state has little scope of taking up major and medium projects. A net area of about 45,000 ha has, so far, been provided with irrigation.

Mizoram

There was hardly any irrigation development in the state in the pre-plan period. The terrain is not suitable for taking up major and medium projects. In the period of plan development, an irrigation potential of 13,000 ha was created through minor irrigation projects, thereby bringing a net area of about 8,000 ha under irrigation. The ultimate irrigation potential of the state is about 70,000 ha, which is one of the lowest of all states.

Nagaland

Nagaland too has a low irrigation potential of about 90,000 ha out of which about 5,000 ha was created in the pre-plan period through minor irrigation schemes. This has been raised to about 68,000 ha by taking up minor irrigation projects. The net area provided with irrigation is of the order of 60,000 ha.

Orissa

Pre-plan development in the state amounted to 0.46 M-ha through 5 major and medium and a large number of minor irrigation projects. During the period of plan development, 55 major and medium projects and a host of minor irrigation projects were undertaken in the state bringing up the irrigation potential created to about 3 M-ha which is more than half the ultimate irrigation potential of 5.9 M-ha for the state. A net area of 2 M-ha has been brought under irrigation through these projects. Important projects taken up during the project include Hirakud, Mahanadi Birupa Barrage, upper Kolab etc. Schemes like interstate Subernarekha project and upper Indravati are presently in progress.

Punjab

Punjab was one of the states where significant development in irrigation was made during pre-plan period. An irrigation potential of about 0.21 M-ha was created in the state through construction of 5 major and a number of minor lift irrigation projects. During the period of plan development 10 major and medium and a large number of minor projects were undertaken which has raised the irrigation potential of 6.5 M-ha. A net area of about 3.9 M-ha has been brought under irrigation in the state out of net sown area of 4.2 M-ha Important major and medium projects undertaken in the state include interstate Bhakra Nangal, Sutlej-Yamuna Link Canal, Beas etc.

Rajasthan

Out of the ultimate irrigation potential of 5.15 M-ha for the state, a potential of 1.5 M-ha was developed during the pre-plan period through the construction of one major and 42 medium irrigation projects in addition to several minor projects. About 67 major and medium projects and a large number of minor projects were undertaken during the plan period which has raised the irrigation potential created to about 4.8 M-ha which is more than 90 percent of ultimate irrigation potential of the state. Net area brought under irrigation is 3.9 M-ha out of net sown area of 16.4 M-ha Important projects undertaken during plan development include interstate Chambal project, Bhakra Nangal, Beas, Indira Nahar, Mahi Bajaj Sagar etc.

Sikkim

There was hardly any irrigation development in Sikkim at the time when plan development started in India. After Sikkim joined as a part of India, plan development was extended to the state in the seventies and an irrigation potential of about 25,000 ha is likely to be developed by the end of 1996-97 through minor irrigation projects.

Tamil Nadu

The state of Tamil Nadu was one of the forerunners in development of irrigation during the British period. An irrigation potential projects of 2.4 M-ha was developed through 24 major and medium irrigation projects and a large number of minor irrigation projects during the pre-plan period. About 1.3 M-ha of irrigation potential was added during plan development through addition of 24 major and medium projects and a number of minor irrigation projects. Out of the ultimate irrigation potential of 3.9 M-ha, over 3.7 M-ha (about 95 percent) has so far been created. Out of the net sown of 5.6 M-ha, net area brought under irrigation is 2.7 M-ha Important projects undertaken during the period include Cauvery Delta, Lower Bhavani, Parambikulam Aliyar etc.

Tripura

Irigation development in the pre-plan period in Tripura was mainly through minor irrigation. Potential of 10,000 ha was created during the pre-plan period. With the addition of 3 medium and several minor. irrigation projects during plan development, the irrigation potential, so far, developed has risen to over 100,000 ha which is nearly half of the ultimate potential of 215,000 ha. About 16,000 ha of sown area have been brought under irrigation. One of the important projects taken up in the state is Gumti.

Uttar Pradesh

With vast development in the Ganga valley an irrigation potential of 5.4 M-ha was created in the state during the pre-plan period, through 15 major schemes and a host of minor schemes. During plan development over 90 major and medium projects and a very large number of minor irrigation projects were undertaken in the plan period thereby raising the potential created to about 30 M-ha against the originally assessed ultimate potential of about 26 M-ha Three-fourth of this development (22.7 M-ha) is attributed to minor irrigation projects- largely ground water works. About 11.3 M-ha of area out of net sown area of 17.3 M-ha has been brought under irrigation through these projects. Important projects undertaken during the period include Ramganga, Sarda Sahayak, Saryu Nahar, Gandak, Madhya Ganga Canal, Tehri etc.

West Bengal

Pre-plan irrigation development in West Bengal saw the implementation of major projects in addition to a large number of minor projects which resulted in creation of an irrigation potential of about 0.94 M-ha During plan period, about 13 major and medium and large number of minor schemes were added which have raised the irrigation potential of 6.1 M-ha A net area of 1.9 M-ha has been brought under irrigation through these projects out of net sown area of 5.3 M-ha Important projects undertaken during the period include DVC Barrage and canal system, Mayurakshi, Kangsabati etc. Since the above information has been based on the available data from Central Water Commission, and Ministry of Water Resources, Government of India for the last decade, the smaller states carved out later have not been included and the data represents that for the undivided state.

Irrigation Development History

The history of irrigation development in India can be traced back to prehistoric times. Vedas and ancient Indian scriptures made reference to wells, canals, tanks and dams which were beneficial to the community and their efficient operation and maintenance was the responsibility of the State. Civilisation flourished on the banks of the rivers and harnessed the water for sustenance of life.

According to the ancient Indian writers, the digging of a tank or well was amongst the greatest of the meritorious act of a man. Brihaspathi, an ancient writer on law and politics, states that the construction and the repair of dams is a pious work and its burden should fall on the shoulders of rich men of the land. Vishnu Purana enjoins merit to a person who patronages repairs to well, gardens and dams. In a monsoon climate and an agrarian economy like India, irrigation has played a major role in the production process. There is evidence of the practice of irrigation since the establishment of settled agriculture during the Indus Valley Civilisation (2500 BC).

These irrigation technologies were in the form of small and minor works, which could be operated by small households to irrigate small patches of land and did not require cooperative effort. Nearly all these irrigation techn-ologies still exist in India with little technological change, and continue to be used by independent households for small holdings.

The lack of evidence of large irrigation works at this time signifies the absence of large surplus that could be invested in bigger schemes or, in other words, the absence of rigid and unequal property rights. While village communities and cooperation in agriculture did exist as seen in well developed townships and economy, such co-operation in the large irrigation works was not needed, as these settlements were on the fertile and well irrigated Indus basin. The spread of agricultural settlements to less fertile and irrigated area led to co-operation in irrigation development and the emergence of larger irrigation works in the form of reservoirs and small canals.

While the construction of small schemes was well within the capability of village communities, large irrigation works were to emerge only with the growth of states, empires and the intervention of the rulers. There used to emerge a close link between irrigation and the state. The king had at his disposal the power to mobilize labour which could be used for irrigation works.

In the south, perennial irrigation may have begun with construction of the Grand Anicut by the Cholas as early as second century to provide irrigation from the Cauvery river. Wherever the topography and terrain permitted, it was an old practice in the region to impound the surface drainage water in tanks or reservoirs by throwing across an earthen dam with a surplus weir, where necessary, to take off excess water, and a sluice at a suitable level to irrigate the land below. Some of the tanks got supplemental supply from stream and river channels. The entire land-scape in the central and southern India is studded with numerous irrigation tanks which have been traced back to many centuries before the beginning of the Christian era. In northern India also there are a number of small canals in the upper valleys of rivers which are very old.

Medieval Period Irrigation

In the medieval India, rapid advances also took place in the construction of inundation canals. Water was blocked by constructing bunds across steams. This raised the water level and canals were constructed to take water to the fields. These bunds were built by both the state and private sources. Ghiyasuddin Tughluq 1220-250 is credited to be the first ruler who encouraged digging canals. However, it is Fruz Tughlug (1351-86) who inspired from central Asian experience, is considered to be the greatest canal builder before the nineteenth century. Irrigation is said to be one of the major reasons for the growth and expansion of the Vijayanagar empire in southern India in the fifteenth century. It may be noted that, but for exceptional cases, most of the canal irrigation prior to the arrival of the British was of the diversionary nature.

The state, through the promotion of irrigation, had sought to enhance revenue and provide patronage through rewards of fertile land and other rights to different classes. Irrigation had also increased employment opportunities and helped in the generation of surplus for the maintenance of the army and the bureaucracy. As agricultural development was the pillar of the economy, irrigation systems were paid special attention to, as irrigation was seen to be a catalyst for enhanced agricultural production. This is demonstrated by the fact that all the large, powerful and stable empires paid attention to irrigation development. It may be said that the state in irrigation was commensurate with its own interest.

Irrigation Development Under British Rule

Irrigation development under British rule began with the renovation, improvement and extension of existing works. When enough experience and confidence had been gained, the Government ventured on new major works, like the Upper

Ganga Canal, the Upper Bari Doab Canal and Krishna and Godavari Delta Systems, which were all river-diversion works of considerable size. The period from 1836 to 1866 marked the investigation, development and completion of these four major works. In 1867, the Government adopted the practice of accepting works, which promised a minimum net return. Thereafter, a number of projects were taken up.

These included major canal works like the Sirhind, the Lower Ganga, the Agra and the Mutha Canals, and the Periyar Dam and canals. Some other major canal projects were also completed on the Indus system during this period. These included the Lower Swat, the Lower Sohag and Para, the Lower Chenab and the Sidhnai Canals, ali of which went to Pakistan in 1947.

The recurrence of drought and famines during the second half of the nineteenth century necessitated the development of irrigation to give protection against the failure of crops and to reduce large scale expenditure on famine relief. As irrigation works in low rainfall tracts were not considered likely to meet the productivity test, they had to be financed from current revenues.

Significant protective works were constructed during the period were the Betwa Canal, the Nira Left Bank Canal, the Gokak Canal, the Khaswad Tank and the Rushikulya Canal. Between the two types of works namely productive and protective, the former received greater attention from the Government. The gross area irrigated in British India by public works at the close of the nineteenth century was about 7.5 M-ha Of this, 4.5 M-ha came from minor works like tanks, inundation canals etc. for which no separate capital accounts were maintained.

The area irrigated by protective works was only a little more than 0.12 M-ha Irrigation development at the time of independence. The net irrigated area in the Indian sub continent, comprising the British Provinces and Princely States, at the time of independence was about 28.2 M-ha, the largest in any country of the world. The partition of the country, however, brought about sudden and drastic changes, resulting in the apportionment of the irrigated area between the two countries; net irrigated area in India and Pakistan being 19.4 and 8.8 M-ha respectively.

Major canal systems, including the Sutlej and Indus systems fell to Pakistan's share. East Bengal, now Bangladesh, which comprises the fertile Ganga Brahmaputra delta region also went to Pakistan. The irrigation works which remained with India barring some of the old works in Uttar Pradesh and in the deltas of the South, were mostly of a protective nature, and meant more to ward off famine than to produce significant yields.

Plan Development

In the initial phase of water resources development during the plan period after independence, rapid harnessing of water resources was the prime objective. Accordingly, the State Governments were encouraged to expeditiously formulate and develop water resources projects for specific purposes like irrigation, flood control, hydro-power generation, drinking water supply, industrial and various miscellaneous uses. As a result, a large number of projects comprising dams, barrages, hydro-power structures, canal net work etc. have come up all over the country in successive Five Year Plans.

A milestone in water resources development in India is creation of a huge storage capability. Because of these created storage works it has now become possible to provide assured irrigation in the command area, to ensure supply for hydro-power and thermal power plants located at different places and to meet requirement for various other uses.

Flood moderation could be effected in flood prone basins, where storage have been provided. Besides, supply of drinking water in remote places throughout the year has become possible in different parts of the country. At the time of commencement of the First Five Year Plan in 1951, popul-ation of India was about 361 million and annual food grain production was 51 million tonnes which was not adequate. Import of food grains was then inevitable to cover up the shortage.

Attaining self sufficiently in food was therefore given paramount importance in the plan period and in order to achieve the objective, various major, medium and minor irrigation and multipurpose projects were formulated and implemented through successive Five Year Plans to create additional irrigation potential throughout the country.

This drive compounded with green revolution in the agricultural sector, has enabled India to become marginally surplus country from a deficit one in food grains. Thus the net irrigated area is 37 percent of net sown are and 29 percent of total cultivable area. The ultimate potential due to major and medium projects has been assessed as 58 M-ha of which 60 per cent estimated to be developed.

Minor irrigation Development Programmes

While the development of irrigation is most essential for increasing food and other agricultural production to meet the needs of the growing population, development of Minor Irrigation should receive greater attention because of the several advantages they posses like small investments, simpler components as also being labour intensive, quick maturing and mot of all farmer friendly.

Minor Irrigation development programmes in the state is being implemented by many Departments/Organisations like Agriculture, Rural Development, Irrigation, Social Welfare etc At the central level also, different departments launch schemes having Minor Irrigation component.

The Ministry of Rural Areas and Employment launched a Million Wells Schemes (WMS) in 1988-89. Till 1997-98, a total of 12,63,090 wells have been constructed under MWS with an expenditure of Rs. 4,728.17 crore. The Ministry of Rural Areas and Employment is also implementing Drought Prone Area Programme(DPAP) on watershed basis. The Ministry of Agriculture has been instrumental in providing credit to farmers for the development of Minor Irrigation through Commercial Banks, Regional Rural Banks, Cooperatives and National Bank for Agriculture and Rural Development (NABARD). The Minor Irrigation Division of the Ministry of

Water Resources monitors the progress of development of irrigation created through Minor Irrigation Project.

It implements the Centrally Sponsored Scheme Rationalisation of Minor Irrigation Statistics (RMIS) and conducts census of Minor Irrigation structures on quinq-uennial basis with a view to create a reliable database for the developm-ent of Minor Irrigation Sector. It also assists the State Governments in preparation of schemes for posing to external funding agencies for attracting external assistance for Minor Irrigation Schemes. Ground Water Development is primarily done through individual and co-operative efforts of the farmers with the help of institutional finance and their own savings.

Surface Water minor Irrigation Scheme i.e. surface lift schemes and surface flow schemes are generally funded from the public sector outlay. NABARD provides finance to the banks for installation of Minor Irrigation works in the States. In addition, the Land Development Banks provide bank credit to the farmers under their normal programmes also. During the year 1997-98, the total credit disbursement for minor irrigation works out of Rs. 488.65 crores. Further, many old schemes go out of use due to one reason or the other. The Irrigation Potential created and utilised through ground water as well as surface water. Minor Irrigation Schemes are not being recorded systematically in most cases as there schemes are implemented and monitored by individual farmers.

Reservoir Storage Position

The storage position in 68 important reservoirs in different parts of the country is monitored by the Central Water Commission. Against the designed live capacity at full reservoir levels of 129.4 Th. Million Cubic meters (TMC) in these reservoirs the total live storage was 95.3 TMC at the end of September, 1999 and 106.3 TMC at the same point of time last year.

Effect of Irrigation Potential

The reassessed Ultimate Irrigation Potential (UIP) is 139.89

million hectare (M-ha). This re-assessment has been done on the basis of the re-assessment of the potential of ground water from 40 M-ha to 64.05 M-ha and re-assessment of potential of surface minor irrigation from 15 M-ha to 17.38 M-ha Thus, there has been an increase of 26.39 M-ha in the UIP of the country, which was 113.5 M-ha Before re-assessment. At the inception of planning in India in 1951 the created irrigation potential was 22.60 M-ha The irrigation potential created upto the end of Eighth Plan has increased to 89.56 M-ha.

Influence of Major and Medium Irrigation

The Ultimate Irrigation Potential of the country from Major and Medium Irrigation projects has been assessed as 58.5 m ha. This includes projects with a culturable command area of more than two thousand hectare. The potential created up to the end of the Seventh Plan (1985-90) was 29.92 M-ha and at the beginning of the Eighth Five Year Plan (1992-93) was 30.74 M-ha A target 5.09 M-ha had been set for creation of additional potential during the Eighth Plan (1992-97) against which, the potential created was about 2.22 M-ha Thus, at the end of the Eighth Plan, the cumulative irrigation potential created from major and Medium irrigation was about 32.96 M-ha According to a provisional estimate, the irrigation potential through Major & Medium projects has reached the level of 34.5 M-ha By 1998-99.

All ground water and surface water schemes having culturable command area (CCA) upto 2000 ha individually are classified as minor irrigation schemes. Ground water development is primarily done through individual and cooperative efforts of the farmers with the help of institutional finance and their own savings. Surface water minor irrigation schemes are generally funded from the public sector outlay.

References

Busman, J.D., *Optimizing control of surface irrigation using concurrent evaluation of infiltration*, PhD Dissertation, Agricultural and Irrigation Engineering, Utah State University, Logan, Utah. Unpublished document, 209p, 1987.

2

Estimation of Irrigation Demand

A plot of land growing a certain crop or a combination of crops has to be supplied with water from time to time. Primarily, the plot or field is expected to receive water from rain falling on the land surface. But the distribution of rain is rather uncertain both in time and space. Also some of the rain as in a light shower does not reach the ground as it may be intercepted by the leaves of the plant during a heavy downpour; much of the water might flow away as surface runoff. Hence, only a certain amount of falling rain may be effective in raising the soil moisture that is actually useful for plant growth. Hence, for proper crop growth, the effective rain has to be supplemented by artificially applying water to the field by irrigation.

If the area of the field is small, water may be supplied from the local ground water source. If the field is large, supplemented irrigation water may be obtained from a local surface water source, like a river, if one is available nearby. The work of a water resources engineer therefore would be to design a suitable source for irrigation after knowing the demand of water from field data.

Water Requirement of a Crop

It is essential to know the water requirement of a crop which is the total quantity of water required from its sowing time up to harvest. Naturally different crops may have different water requirements at different places of the same country, depending upon the climate, type of soil, method of cultivation, effective rain etc.

The total water required for crop growth is not uniformly distributed over its entire life span which is also called crop period. Actually, the watering stops same time before harvest and the time duration from the first irrigation during sowing up to the last before harvest is called base period. Though crop period is slightly more than the base period, they do not differ from practical purposes.

Sometimes, in the initial stages before the crop is sown, the land is very dry. In such cases, the soil is moistened with water as to helps in sowing the crops. This is known as paleo irrigation. A term kor watering is used to describe the watering given to a crop when the plants are still young. It is usually the maximum single watering required, and other waterings are done at usual intervals. The total depth of water required to raise a crop over a unit area of land is usually called delta.

Duty of Water

The term duty means the area of land that can be irrigated with unit volume of irrigation water. Quantitatively, duty is defined as the area of land expressed in hectares that can be irrigated with unit discharge, that is, 1 cumec flowing throughout the base period, expressed in days. Imagine a field growing a single crop having a base period B days and a Delta Δ mm which is being supplied by a source located at the head (uppermost point) of the field. The water being supplied may be through the diversion of river water through a canal, or it could be using ground water by pumping (Figure 1). If the water supplied is just enough to raise the crop within D hectares of the field, then a relationship may be found out amongst all the variables as:

Volume of water supplied = $B*60*60*24\ m^3$

Area of crop irrigated = $D*10^4\ m^2$

Volume of water supplied per unit area = $\frac{86400}{10000\,D} = \frac{8.64\,B}{D}$

Hence, knowing two of the three variables B, D and Δ the third party may be found out.

Figure 1. Water applied to field by pumbing ground water

The duty of irrigation water depends upon a number of factors; some of the important ones are as follows:

— *Type of crop*: As different crops require different amount of water for maturity, duties are also required. The duty would vary inversely as the water requirement of crop.

— *Climate season and type of soil*: Some water applied to the field is expected to be lost through evaporation and deep percolation. Evaporation loss has a direct bearing on the prevalent climate and percolation may be during drier seasons when the water table is low and soil is also dry. Percolation loss would be more for sandy soils than silty or clayey soils.

— *Efficiency of cultivation methods*: If the tillage and methods of water application are faulty and less efficient, then the amount of water actually reaching the plant roots would be less. Hence, for proper crop growth more water would be required than an equivalent efficient system. Also, if the water is conveyed over long distances through field channels before being finally applied to the field, then also the duty will rise due to the losses taking place in the channels.

Crop Growing Seasons in India

Each crop has its own sowing and harvesting seasons and it is important to have a knowledge of this which may help to decide the total water demand in a field having mixed crops. In India, the northern and north eastern regions have two distinct cropping seasons. The first coinciding mostly with the South western monsoon is called kharif, which spans mostly from July to October. The other, called rabi, spans generally over October to March. The summer season crops are planted sometime between April and June. In southern part of India, there is no such distinct season, but each region has its own classification of seasons.

Generally, the kharif is characterised by a gradual fall in temperature, more numerous cloudy days, low intensity, high relative humidity and cyclonic weather. During Rabi, there is a gradual rise in temperature, bright sunshine, near absence of cloud days, and a lower relative humidity.

Country's Irrigation Demands

It may be appreciated that in India there is a large variation of rainfall, which is the primary source of irrigation in most parts of the country. In fact, the crops grown in various regions have been adapted according to the local rainfall availability. Water resources engineers are therefore concerned with arranging supplementary water to support the crops for seasonal variations of rainfall in order to ensure an assured crop harvest.

Cropping Patterns

Planning of an irrigation project requires estimation of water demand of a cultivated area. Naturally, this would depend upon the type of crop grown. Since irrigation water may have to be supplied to one field growing a combination of crops or to many fields growing different crops, it is important to understand certain cropping practices which would be helpful in estimating the irrigation demand. Some of the prevalent practices are as follows:

- Crops grown solely or mixed: Mixed cropping
- Crops grown in a definite sequence: Rotational cropping
- Land occupied by one crop during one season: Mono cropping
- Land occupied by two crops: double cropping
- Land sowed with more than one crop in a year: multiple cropping

For raising a field crop effectively, it is essential to supply water through artificial irrigation supplementing the rain falling over the plot of land and raising the soil moisture. Hence, it may be seen that irrigation water requirement is rather a dynamic one. Also, the crop water requirement is shown with slight variation, it actually shows more variation, depending on the type of crop and the prevalent climate. Though farmers may be tempted to allow more water to the plants through supplemental irrigation, it must be remembered that there is an optimum water requirement schedule of each crop depending upon its stage of growth. It has been proved that at times application of more water may cause reduction in yield.

The total water need for various plants, known as delta. However, in planning the supply of irrigation water to a field crop, it is essential to estimate the water requirement of each plot of land growing a crop or crops at any point of time. This may be done by studying the dynamic interaction between a crop and the prevalent climate and the consequent water requirement. The demand would, naturally be also dependant on the type of crop and its stage of growth.

Plant roots extract water from the soil. Most of this water doesn't remain in the plant, but escapes to the atmosphere as vapour through the plants leaves and stems, a process which is called transpiration and occurs mostly during daytime. The water on the soil surface as well as the water attaching to the leaves and stem of a plant during a rainfall also is lost to the atmosphere by evaporation. Hence, the water need of a crop consists of transpiration plus evaporation, together called evapotranspiration. The effect of the major climatic factors on crop water needs may be summarised as follows:

— Sunshine

— Temperature

— Humidity

— Wind speed

Since the same crop grown in different climatic variations have different water needs, it has been accepted to evaluate the evapotranspiration rate for a standard or reference crop and find out that of all other crops in terms of this reference. Grass has been chosen as standard reference for this purpose.

The evapotranspiration rate of this standard grass is, therefore, called the reference crop evapotranspiration and is denoted as ET_o, which is of course, the function of the climatic variables. Training Manual 3: Irrigation Water Needs published by the Food and Agricultural Organisation, (FAO) and available on-line through the under-mentioned web-site gives an idea about the variation of ET_o under different climatic conditions and is reproduced in the table 1.

Table 1. showing the daily variation of water needs of standard grass (in mm) under different climatic patterns (ET_o)

	Mean daily Temperature		
Climatic Zone	Low (<150C)	Medium (150-250C)	High (>250C)
Desert/Arid	4-6	7-8	9-10
Semi-arid	4-5	6-7	8-9
Sub-humid	3-4	5-6	7-80
humid	1-2	3-4	5-6

Agricultural scientists have evaluated a factor called crop factor and denoted it by K_C, to evaluate specific crop water needs. Naturally, K_c would be different for different crops and would not be the same throughout the growth season of one type of crop. Thus, the crop evapotranspiration, denoted by ET_C is to be evaluated as under:

$$ET_O = K_C * ET_C$$

Both ET_O and ET_C should be in the same units and generally, mm/day is used as a standard all over the world. In order to simply the calculations, the factor K_C has been evaluated for 4 stages of a crop growth usually denoted as

— Initial stage

— Crop development stage

— Mid-season stage

— Late season stage

The FAO Training Manual 3 gives the growth stage periods and the corresponding K_C values for some typical crops. In the table 2, that for rice is presented.

Table 2

Rice	Climate			
	Little wind		Strong wind	
Growth stage	Dry	Humid	Dry	Humid
0-60 days	1.1	1.1	1.1	1.1
Mid season	1.2	1.05	1.35	1.3
Last 30 days before harvest	1.0	1.0	1.0	1.0

It may be mentioned that any crop doesn't have a fixed total growth period, which is the summation of growth stage periods given above. There is usually a range, depending upon the variety of the crop and the condition in which it is cultivated. The values of K_C also depend upon the climate and particularly on humidity and wind speed, as shown for rice in the above table.

In general, the values of K_C should be reduced by 0.05 if the relative humidity is high (>80%) and the wind speed is low (<2m/s). Likewise, the values should be increased by 0.05 if the relative humidity is low (<50%) and the wind speed is high (>5m/s). For full details, the FAO training manual 3 may be consulted as K_C values for other crops are evaluated in different manners.

Reference Crop ET_O

Of the many methods available, the commonly used ones are two:

— Experimental methods, using the experimentation data from evaporation pan.

— Theoretical methods using empirical formulae, that take into account, climatic parameters.

Experimental Method

Estimation of ET_O can be made using the formula

$$ET_O = Kpan \times Epan$$

Where ETO is the reference crop evapotranspiration in mm/day, Kpan is a coefficient called pan coefficient and Epan is the evaporation in mm/day from the pan. The factor Kpan varies with the position of the equipment, humidity and wind speed. Generally, the details are supplied by the manufacturers of the pan. For the US Class A evaporation pan, which is also used in India, Kpan varies between 0.35 and 0.85, with an average value of 0.7.

It may be noticed that finding out ET_C would involve the following expression

$$ET_C = Kcrop \times ET_O = Kc \times Epan \times Kpan$$

If instead, Kcrop x Kpan is taken as a single factor, say K, then ET_C may directly be found from Epan as under:

ETC =K x Epan, where K may be called the crop factor

The water management division of the Department of Agriculture, Government of India has published a list of factors for common crops and depending upon the stage of growth, which have to be multiplied with the evaporation values of the USWB Class A evaporation pan.

Theoretical Methods

The important methods that have been proposed over the years take into account, various climatic parameters.

Blanney-Criddle formula: This formula gives an estimate of the mean monthly values of ET_O which is stated as

$$ET_O = p\ (0.46\ Tmean + 8.13)$$

Where p is the mean daily percentage of annual day time hours and has been estimated according to latitude; Tmean is the mean monthly temperature in degrees Centigrade and may be taken as ½ x (Tmax + Tmin) for a particular month.

Penman-Monteith method: This method suggests that the value of ET_O may be evaluated by the following formula:

$$ET_O = \frac{0.408\,\Delta(R_n - G) + \gamma \dfrac{900}{T+273} u_2 (e_s - e_a)}{\Delta + \gamma(1 + 0.34u_2)}$$

Where the variables have the following meanings:

ET_O reference evapotranspiration [mm day^{-1}],

R_n net radiation at the crop surface [MJ m^{-2} day^{-1}],

G soil heat flux density [MJ m^{-2} day^{-1}],

To mean daily air temperature at 2 m height [°C],

u_2 wind speed at 2 m height [m s^{-1}],

e_s saturation vapour pressure [kPa],

e_a actual vapour pressure [kPa],

e_s - e_a saturation vapour pressure deficit [kPa],

D slope vapour pressure curve [kPa °C^{-1}],

g psychrometric constant [kPa °C^{-1}].

Application Interval of Irrigation Water

The water need of a crop is usually expressed as mm/day, mm/month or mm/season, where season means the crop growing period. Whatever be the water need, it need not be applied each day. A larger amount of water may be applied once in a few days and it gets stored in the crop root zone, from where the plant keeps on extracting water. Soon after irrigation, when the soil is saturated, up to the field capacity, the extraction of water from the soil by the plants is at the peak. This rate of water withdrawal decreases as the soil moisture depletes.

A stage is reached, in the moisture content of the soil, below which the plant is stressed to extract and unless the soil moisture is increased by application of water, the plant production would decrease. The difference of moisture content between field capacity (the maximum content of available water) and the lowest allowable moisture content is called the optimum soil water. The optimum soil moisture range for some common crops is required from which the interval period of irrigation water may be estimated as follows:

$$Irrigation\ period\ (days) = \frac{\begin{array}{c}Net\ depth\ of\ soil\ depletion\ in\ the\ crop\\ area\ just\ before\ irrigation\,(mm)\end{array}}{ETc\ (mm/day)}$$

Where the crop evapotranspiration rate (ET_C) may be determined according to the crop type. The irrigation period, has not taken the soil retention characteristics. Naturally, a soil with greater water retentive capacity serves as a bigger water reservoir for crops and supply of irrigation can be delayed. Consequently, frequency of irrigation is lower and interval of irrigation is longer in heavier soils and in soils with good organic content and low content of soluble salts.

Total Water Requirement in Growing a Crop

The water that is required to irrigate a field or plot of land growing the particular crop not only has to satisfy the evapotranspiration needs for growing the crop, but would also include the following:

- Losses in the form of deep percolation while conveying water from the inlet of the field upto its last or tail end as the water gets distributed within the field
- Water requirement for special operations like land preparation, transplanting, leaching of salts, etc.

The evapotranspiration requirement of crops (ET) really doesn't include the water required by crops for building up plant tissues, which is rather negligible compared to the evaporation needs. Hence ET_C is often equivalently taken as the consumptive irrigation requirement (CIR).

The net irrigation requirement (NIR) is defined as the amount of irrigation water required to be delivered in the field to meet the consumptive requirement of crop as well as other needs such as leaching, pre-sowing and nursery water requirement (if any). Thus,

$$NIR = CIR + LR + PSR + NWR$$

Where

LR = Leaching requirement

PSR = Pre-sowing requirement

NWR = Nursery water requirement

Field Irrigation Requirement (FIR) is defined as the amount of water required to meet the net irrigation requirements plus the amount of water lost as surface runoff and through deep percolation. Considering a factor η_a called the water application efficiency or the field application efficiency which accounts for the loss of irrigation water during its application over the field, we have

$$FIR = \frac{NIR}{\eta_a}$$

Now, consider an irrigated area where there is a single source of water is supplying water to a number of fields and water is applied to each field by rotation. Naturally, some water is lost through the respective turnouts. Hence, the source must supply a larger amount of water than that required at any point of time by adding up the flows to the fields turnouts that are open at that point of time. Thus, the capacity of the water supply source may be termed as the gross irrigation requirement (GIR), defined as:

$$GIR = \frac{FIR}{\eta_c}$$

In the above equation, η_c is the water conveyance efficiency.

Sometimes, in the initial stages before the crop is sown, the land is very dry. This happens usually at the time of sowing Rabi crops because of hot September, when the soil may be too dry to be sown easily. In such a case, the soil is first moistured with water to help to sowing of seeds, and the water application for this purpose is known as Paleo Irrigation.

The total quality of water required by a crop is applied through a number of waterings at certain intervals throughout the base period of the crop. However, the quantity of water required to be applied during each of these waterings is not the same. In general, for all crops during the first watering after the plants have grown a few centimetres high, the quantity of water required is more than that during subsequent waterings. The first watering after the plants have grown a few centimetres high is known as Kor watering and the depth of water applied during watering is known as Kor depth. The watering must be done in a limited period which is known as the Kor period. The duty of water at the outlet that is at the turnout leading from the water courses and field channels on the field is known as the outlet factor.

References

Dilip Kumar Majumdar, *Irrigation Water Management (Principles & Practices)*, Prentice Hall of India (P) Ltd.

3

Irrigation Schemes and Methods

Irrigation projects are classified in different ways, however, in Indian context it is usually classified as follows:

- *Major project*: This type of project consists of huge surface water, storage reservoirs and flow diversion structures. The area envisaged to be covered under irrigation is of the order over 10000 hectare.
- *Medium project*: These are also surface water projects but with medium size storage and diversion structures with the area under irrigation between 10000 hectare and 2000 hectare.
- *Minor project*: The area proposed under irrigation for these schemes is below 2000Ha and the source of water is either ground water or from wells or tube wells or surface water lified by pumps or by gravity flow from tanks. It could also be irrigated from through water from tanks.

The major and medium irrigation projects are further classified as

- Direct irrigation method
- Storage irrigation method.

Commanded area (CA): is defined as the area that can be irrigated by a canal system, the CA may further be classified as under:

Gross command area (GCA): This is defined as total area that can be irrigated by a canal system on the perception that unlimited quantity of water is available. It is the total area that may theoretically be served by the irrigation system. But this may include inhibited areas, roads, ponds, uncultivable areas etc which would not be irrigated.

Culturable command area (CCA): This is the actually irrigated area within the GCA. However, the entire CCA is never put under cultivation during any crop season due to the following reasons:

- The required quantity of water, fertilizer, etc. may not be available to cultivate the entire CCA at a particular point of time. Thus, this is a physical constraint.
- The land may be kept fallow that is without cultivation for one or more crop seasons to increase the fertility of the soil. This is a cultural decision.
- Due to high water table in some areas of the CCA irrigated water may not be applied as the crops get enough water from the saturation provide to the surface water table.

During any crop season, only a part of the CCA is put under cultivation and this area is termed as culturable cultivated area. The remaining area which is not cultivated during a crop season is conversely termed as culturable uncultivated area. Intensity of irrigation is defined as the percentage of the irrigation proposed to be irrigated annually. Usually the areas irrigated during each crop season (Rabi, Kharif, etc) is expressed as a percentage of the CCA which represents the intensity of irrigation for the crop season. By adding the intensities of irrigation for all crop seasons the yearly intensity of irrigation to be obtained. As such, the projects with a CCA of more than 2000 hectare are grouped as major and medium irrigation projects. The ultimate irrigation potential of our country from major and medium projects has been assessed as 58.46 M-hectare.

As per the report of the Ministry of Water Resources, Government of India the plan wise progress of irrigation of creation of irrigation potential through major and medium projects is as follows. The Table 1 provides a list of the Major

Table 1: Areas given in thousand hectares

Sl. No.	State Name	Potential created upto the end of VIII Plan (1992-97)	Potential utilised upto the end of VIII Plan (1992-97)
1.	Andhra Pradesh	3045.10	2883.80
2.	Arunachal Pradesh	-	-
3.	Assam	196.67	138.17
4.	Bihar	2802.50	2324.20
5.	Goa	13.02	12.07
6.	Gujarat	1350.00	1200.00
7.	Haryana	2078.79	1833.62
8.	Himachal Pradesh	10.55	5.59
9.	Jammu & Kashmir	173.70	147.57
10.	Karnataka	1666.02	1471.70
11.	Kerala	513.31	464.31
12.	Madhya Pradesh	2317.60	1620.95
13.	Maharashtra	2313.00	1287.70
14.	Manipur	63.00	52.00
15.	Meghalaya	-	-
16.	Mizoram	-	-
17.	Nagaland	-	-
18.	Orissa	1557.75	1442.66
19.	Punjab	2512.85	2452.34
20.	Rajasthan	2273.88	2088.39
21.	Sikkim	-	-
22.	Tamil Nadu	1545.51	1545.49
23.	Tripura	2.30	2.30
24.	Uttar Pradesh	7059.00	6126.00
25.	West Bengal	1444.08	1332.52
Total - States		32938.63	28431.38
Total - UT		18.51	9.29
Grand Total		32957.14	28440.67

and Medium irrigation state-wise creation and utilisation of irrigation potential at the end of VIII plan (1992-97).

Table 2: Areas given in thousand hectares

Sl. No.	State Name	Net Sown Area (NSA)	Net Irrigated Area (NIA)	% of NIA to NSA
1.	Andhra Pradesh	10637	4123	38.76
2.	Arunachal Pradesh	185	36	19.46
3.	Assam	2780	572	20.57
4.	Bihar	7321	3680	50.27
5.	Goa	139	23	16.55
6.	Gujarat	9609	3002	31.24
7.	Haryana	2586	2761	76.99
8.	Himachal Pradesh	568	101	17.78
9.	Jammu & Kashmir	734	386	52.59
10.	Karnataka	10420	2302	22.09
11.	Kerala	2265	342	15.10
12.	Madhya Pradesh	19752	5928	30.01
13.	Maharashtra	17911	2567	14.33
14.	Manipur	140	65	46.43
15.	Meghalaya	206	45	21.84
16.	Mizoram	109	7	6.42
17.	Nagaland	211	62	29.38
18.	Orissa	6210	2090	33.65
19.	Punjab	4139	3847	92.94
20.	Rajasthan	16575	5232	31.56
21.	Sikkim	95	16	16.84
22.	Tamil Nadu	5342	2625	49.14
23.	Tripura	277	35	12.64
24.	Uttar Pradesh	17399	11675	67.10
25.	West Bengal	5462	1911	34.99
Total - States		142072	53433	37.61
Total - UT		143	75	52.45
Grand Total		142215	53508	37.62

It may be noted that the information contained in the table does not separate out the newly formed states which were created after 1997. As for the minor irrigation schemes mostly using ground water sources are primarily developed through individual and cooperative efforts of the farmers with the help of institutional finance and their own savings. Surface water minor irrigation schemes (like lifting water by pumps from rivers) are generally funded from the public sector outlay. The ultimate irrigation potential from minor irrigation schemes has been assessed as 81.43 hectare. The development of minor irrigation should receive greater attention because of the several advantages they possess like small investments, simpler components has also being labour incentive quick money and most of all farmers friendly.

The importance of irrigation in the Indian agricultural economy be appreciated at a glance of the following table showing state wise details of net sown areas and the area that is irrigated (net irrigated areas) for states like Punjab the area irrigated is more than 90 percent followed by Haryana (77 percent) and Uttar Pradesh (67 percent). The national average is low, at around 38 percent.

Direct and Indirect Irrigation Methods

The major and medium surface water schemes are usually classified as either direct or indirect irrigation projects and these are defined as follows:

Direct Irrigation Method

In this project water is directly diverted from the river into the canal by constructing a diversion structure like weir or barrage across the river with some poundage to take care of diurnal variations. It also effects in raising the river water level which is then able to flow into the offtaking channel by gravity. The flow in the channel is usually controlled by a gated structure and this in combination with the diversion structure is also sometimes called the headworks. If the water from such headworks is available throughout the period of growth of crops irrigated by it, it is called a perennial irrigation scheme. In this type of projects, the water in the offtaking channels from the river carries water through out the year. It may not be necessary, however, to provide irrigation water to the fields during monsoon.

In some places local rainfall would be sufficient to meet the plant water needs. In case of a non-perennial river the offtaking channel would be carrying water only for certain period in a year depending upon the availability of supply from the source. Another form of direct irrigation is the inundation irrigation which may be called river-canal irrigation. In this type of irrigation there is no irrigation work across the river to control the level of water in the river. Inundation canal off-taking from a river is a seasonal canal which conveys water as and when available in the river.

This type of direct irrigation is usually practiced in deltaic tract that is, in areas having even and plane topography. It is feasible when the normal flow of river or stream throughout the period of growth of crop irrigated, is never less than the requirements of the irrigated crops at any time of the base period. The diversion structure raises the river water level and is just sufficient to force some water into the channel, the stored water in the pond created behind doesn't have sufficient storage volume it may however be able to take care of any diurnal variation in the river water.

Storage Irrigation Method

For this type of irrigation schemes part of the excess water of a river during monsoon which other wise would have passed down the river as a flood is stored in a reservoir or tank found at the upstream of a dam constructed across a river or stream. This stored water is then used for irrigation is adopted when the flow of river or stream is in excess of the requirements of irrigated crops during a certain part of the year but falls below requirements or is not available at all in the river during remaining part of the year. Since the construction site of a

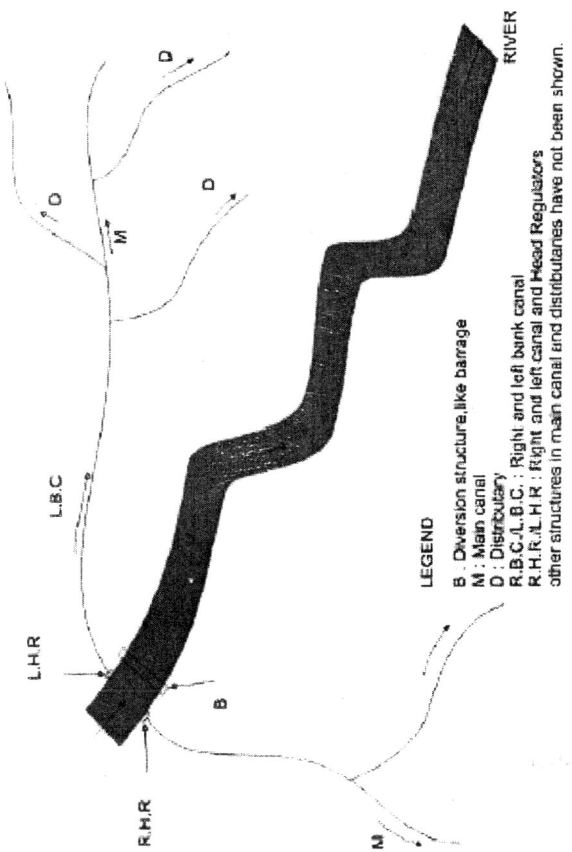

Figure 1. An example of a direct irrigation method

storage reservoir is possible in regions of undulating topography, it is usually practiced in non deltaic areas.

In third type of scheme the storage head works or the dams has to be equipped with ancillary structure like outlet, sluice, spillway, log chutes, etc. The storage created by the dam behind the reservoir is substantial compared to that behind a barrage and may inundate a large tract of land, depending on the topography. The capacity of the reservoir is generally determined systematically by knowing possible withdrawal demands over the weeks and months of a year and corresponding expected inflows.

Another type of storage irrigation method envisages the storage of water at some place in the hilly terrain of the river where the construction of the dam is possible. A barrage is constructed at some downstream location, where the terrain is flatter and canals take off as in a usual direct irrigation method.

Irrigation Project Structures

A number of structures are required for the successful implementation of a project. Some of these are:

Storage structure and appurtenant works

- Dams
- Spillways and energy dissipators
- Sluices and outlets

Diversion structure and appurtenant works

- Barrage
- Canal head regulator
- River training works

Canal water conveyance structures

- Canal sections and layout
- Cross regulators
- Drops
- Turnouts

Also, for an irrigation scheme to be successful, it is important that only the right amount of water be applied to the fields at any point of time. Hence, excess water carried by the canals, as during rainy season, needs to be removed through a drainage network of channels and returned back to the river. Construction of such drainage channels form an important part of a properly designed irrigation network. The Ultimate aim of irrigation planning is to distribute the water meant for irrigation that is stored in storage reservoirs, diverted through diversion structures, conveyed through canal network and discharged to the fields through turnouts to be properly distributed throughout the field. Depending on the type of crop to be grown, the terrain topography and soil characteristics, climate and other local factors different ways and means have been evolved for field water application.

Methods of Field Water Application

Irrigation water conveyed to the head or upstream point of a field must be applied efficiently on the whole area such that the crops growing in the either fields gets water more or less uniformly. Naturally it may be observed that a lot depends on the topography of the land since a large area with uneven topography would result in the water spreading to the low lying areas. The type of crop grown also immensely matter as some like rice, require standing water depths at almost all stages of its growth.

Some, like potato, on the other hand, suffer under excess water conditions and require only the right amount of water to be applied at the right time. Another important factor determining the way water is to apply in the fields is the quantity of water available at any point of time. If water is scarce, as what is actually happening in many parts of the country, then it is to be applied through carefully controlled methods with minimum amount of wastage. Usually these methods employ pressurised flow through pipes which is either sprinkled over the crop or applied carefully near the plant root.

On the other hand when water is rather unlimited during the crop growing season as in deltaic regions, the river flood water is allowed to inundate as much area as possible as long the excess water is available. Another important parameter dictating the choice of the irrigation method is the type of soil. Sometimes water is applied not on the surface of the field but is used to moist the root zone of the plants from beneath the soil surface. Thus, in effective the type of irrigation methods can be broadly divided as under:

— Surface irrigation method

— Subsurface irrigation method

— Sprinkler irrigation system

— Drip irrigation system

Effect of Surface Irrigation Methods

In this system of field water application the water is applied directly to the soil from a channel located at the upper reach of the field. It is essential in these methods to construct designed water distribution systems to provide adequate control of water to the fields and proper land preparation to permit uniform distribution of water over the field. One of the surface irrigation method is flooding method where the water is allowed to cover the surface of land in a continuous sheet of water with the depth of applied water just sufficient to allow the field to absorb the right amount of water needed to raise the soil moisture up to field capacity,.

A properly designed size of irrigation stream aims at proper balance against the intake rate of soil, the total depth of water to be stored in the root zone and the area to be covered giving a reasonably uniform saturation of soil over the entire field. Flooding method has been used in India for generations without any control what so ever and is called uncontrolled flooding. The water is made to enter the fields bordering rivers during folds.

When the flood water inundates the flood plane areas, the water distribution is quite uneven, hence not very efficient, as a lot of water is likely to be wasted as well as soils of excessive slopes are prone to erosion. However the adaptation of this method doesn't cost much. The flooding method applied in a controlled way is used in two types of irrigation methods as under:

— Border irrigation method

— Basin irrigation method

As the names suggest the water applied to the fields by this inundates or floods the land, even if temporarily. On the other hand there are many crops which would try better if water is applied only near their root zone instead of inundating. Such an irrigation method is called the Furrow irrigation method

Features of Border Irrigation

The essential feature of the border irrigation is to provide an even surface over which the water can flow down the slope with a nearly uniform depth. The water spreads and flow down the strip in a sheet confined by border ridges. When the advancing water reaches the lower end of the border, the stream is turned off. For uniform advancement of water front the borders must be properly levelled.

The straight border irrigation is generally suited to the larger mechanized farms as it is designed to produce long uninterrupted field lengths for ease of machine operations. Borders can be 800m or more in length and 3 - 30 m wide depending on variety of factors. It is less suited to small scale farms involving hand labour or animal powered cultivation methods. Generally, border slopes should be uniform, with a minimum slope of 0.05% to provide adequate drainage and a maximum slope of 2% to limit problems of soil erosion.

As for the type of soil suitable for border irrigation, deep homogeneous loam or clay soils with medium infiltration rates are preferred. Heavy, clay soils can be difficult to irrigate with border irrigation because of the time needed to infiltrate sufficient water into the soil. Basin irrigation is preferable in such circumstances.

Basin Irrigation Method

Basins are flat areas of land surrounded by low bunds. The bunds prevent the water from flowing to the adjacent fields. The basins are filled to desired depth and the water is retained until it infiltrates into the soil. Water may be maintained for considerable periods of time. Basin method of irrigation can be formally divided into two, viz; the check basin method and the ring basin method. The check basin method is the most common method of irrigation used in India. In this method, the land to be irrigated is divided into small plots or basins surrounded by checks, levees (low bunds). Each plot or basin has a nearly level surface.

The irrigation water is applied by filling the plots with water up to the desired depth without overtopping the levees and the water retained there is allowed to infiltrate into the soil. The levees may be constructed for temporary use or may be semi permanent for repeated use as for paddy cultivation. The size of the levees depends on the depths of water to be impounded as on the stability of the soil when wet. Water is conveyed to the cluster of check basins by a system of supply channels and lateral field channels or ditches. The supply channel is aligned on the upper side of the field for every two rows of plot. The size of basins depends not only on the slope but also on the soil type and the available water flow to the basins. Generally, it is found that the following holds good for basin sizes.

Basin size should be small if the

— Slope of the land is steep.

— Soil is sandy.

— Stream size to basin is small.

— Required depth of irrigation application is small.

— Field preparation is done by hand or animal traction

Basin size can be large if the

— Slope of the land is flat

— Soil is clay.

— Stream size to the basin is large
— Required depth of the irrigation is large.
— Field preparation is mechanised.

Basin irrigation is suitable for many field crops. Paddy rice grows best when its roots are submerged in water and so basin irrigation is the best method for use with the crop. The other form of basin irrigation is the ring basin method which is used for growing trees in orchards. In this method, generally for each tree, a separate basin is made which is usually circular in shape. Sometimes, basin sizes are made larger to include two more trees in one basin. Water to the basins is supplied from a supply channel through small field channels conveyed the basins with the supply channel. Trees which can be irrigated successfully using the ring basin method include citrus and banana.

Basins can also be constructed on hillside. Here, the ridges of the basins are constructed as in contour border method thus making the only difference between the two is in the application of water. In the border method, the water is applied once during an irrigation cycle and is allowed to flow along the field and as the water infiltrates, till the supply is cutoff. In the basin method, as in a rice field the water is higher at a desired level on the basin. Basin irrigation is suitable for many field crops. Paddy rice grows best when its roots are submerged in water and so basin irrigation is the best method for use with this crop.

Furrow Irrigation

Furrows are small channels, which carry water down the land slope between the crop rows. Water infiltrates into the soil as it moves along the slope. The crop is usually grown on ridges between the furrows. This method is suitable for all row crops and for crops that cannot stand water for long periods, like 12 to 24 hours, as is generally encountered in the border or basin methods of irrigation. Water is applied to the furrows by letting in water from the supply channel, either by pipe siphons or by making temporary breaches in the supply channel embankment. The length of time the water is to flow in the furrows depends on the amount of water required to replenish the root zone and the infiltration rate of the soil and the rate of lateral spread of water in the soil.

Furrow irrigation is suitable to most soils except sandy soils that have very high infiltration water and provide poor lateral distribution water between furrows. As compared to the other methods of surface irrigation, the furrow method is advantageous as:

- Water in the furrows contacts only one half to one-fifth of the land surface, thus reducing puddling and clustering of soils and excessive evaporation of water.
- Earlier cultivation is possible

Furrows may be straight laid along the land slope, if the slope of the land is small (about 5 percent) for lands with larger slopes, the furrows can be laid along the contours.

Subsurface Irrigation Methods

As suggested by the name, the application of water to fields in this type of irrigation system is below the ground surface so that it is supplied directly to the root zone of the plants. The main advantages of these types of irrigation is reduction of evaporation losses and less hindrance to cultivation works which takes place on the surface.

There may be two ways by which irrigation water may be applied below ground and these are termed as:

- Natural sub-surface irrigation method
- Artificial sub-surface irrigation method

Natural Sub-surface irrigation method

Under favourable conditions of topography and soil conditions, the water table may be close enough to the root zone of the field of crops which gets its moisture due to the upward capillary movement of water from the water table.

The natural presence of the water table may not be able to supply the requisite water throughout the crop growing season. However, it may be done artificially by constructing deep channels in the field which may be filled with water at all times to ensure the presence of water table at a desired elevation below the root zone depth. Though this method of irrigation is excellent from both water distribution and labour saving points of view, it is favourable mostly for the following

- The soil in the root zone should be quite permeable
- There should be an impermeable substratum below the water table to prevent deep percolation of water.
- There must be abundant supply of quality water that is one which is salt free, otherwise there are chances of upward movement of these salts along with the moisture likely to lead the conditions of salt incrustation on the surface.

The concept of maintaining a suitable water table just below the root zone is obtained by providing perforated pipes laid in a network pattern below the soil surface at a desired depth. This method of irrigation will function only if the soil in the root zone has high horizontal permeability to permit free lateral movement of water and low vertical permeability to prevent deep percolation of water. For uniform distribution of water percolating into the soil, the pipes are required to be very closely spaced, say at about 0.5m.

Sprinkler Irrigation

Sprinkler irrigation is a method of applying water which is similar to natural rainfall but spread uniformly over the land surface just when needed and at a rate less than the infiltration rate of the soil so as to avoid surface runoff from irrigation. This is achieved by distributing water through a system of pipes usually by pumping which is then sprayed into the air through sprinklers so that it breaks up into small water drops which fall to the ground.

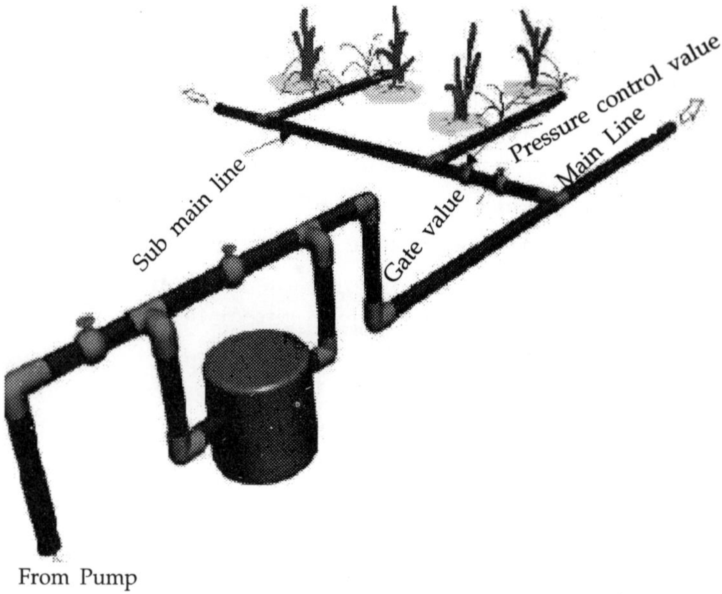

Figure 2. Sprinkler irrigation system

The system of irrigation is suitable for undulating lands, with poor water availability, sandy or shallow soils, or where uniform application of water is desired. No land levelling is required as with the surface irrigation methods. Sprinklers are, however, not suitable for soils which easily form a crust. The water that is pumped through the pump pipe sprinkler system must be free of suspended sediments. As otherwise there would be chances of blockage of the sprinkler nozzles.

A typical sprinkler irrigation system consists of the following components:

— Pump unit

— Mainline and sometimes sub mainlines

— Laterals

— Sprinklers

Figure 2 shows a typical layout of a sprinkler irrigation system. The pump unit is usually a centrifugal pump which takes water from a source and provides adequate pressure for delivery into the pipe system. The mainline and sub mainlines are pipes which deliver water from the pump to the laterals. In some cases, these pipelines are permanent and are laid on the soil surface or buried below ground.

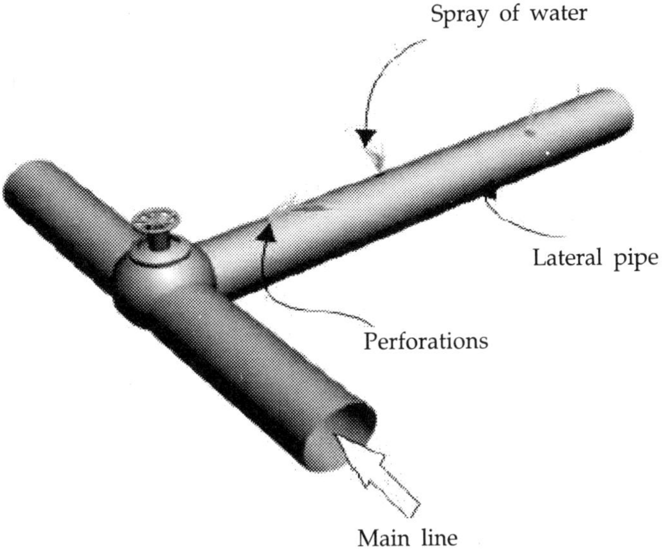

Figure 3. Perforated pipe type of sprinkler system

In other cases, they are temporary, and can be moved from field to field. The main pipe materials include asbestos cement, plastic or aluminum alloy. The laterals deliver water from the mainlines or sub mainlines to the sprinklers. They can be permanent but more often they are portable and made of aluminium alloy or plastic so that they can be moved easily.

The most common types of sprinklers that are used are:

— *Perforated pipe system*: This consists of holes perforated in the lateral irrigation pipes in specially designed pattern to distribute water fairly uniformly. The sprays emanating from the perforations are directed in both sided of the pipe and can cover a strip of land 6 m to 15m wide.

— *Rotating head system*: Here small sized nozzles are placed on riser pipes fixed at uniform intervals along the length of the lateral pipe. The lateral pipes are usually laid on the ground surface. The nozzle of the sprinkler rotates due to a small mechanical arrangement which utilizes the thrust of the issuing water.

As such, sprinkler irrigation is suited for most rows, field as tree crops and water can be sprayed over or under the crop canopy. However, large sprinklers are not recommended for irrigation of delicate crops such as lettuce because the large water drops produced by the sprinklers may damage the crop. Sprinkler irrigation has high efficiency. It however, varies according to climatic conditions; 60% in warm climate; 70% in moderate climate and 80% in humid or cool climate. Sprinkler irrigation was not widely used in India before the 1980.

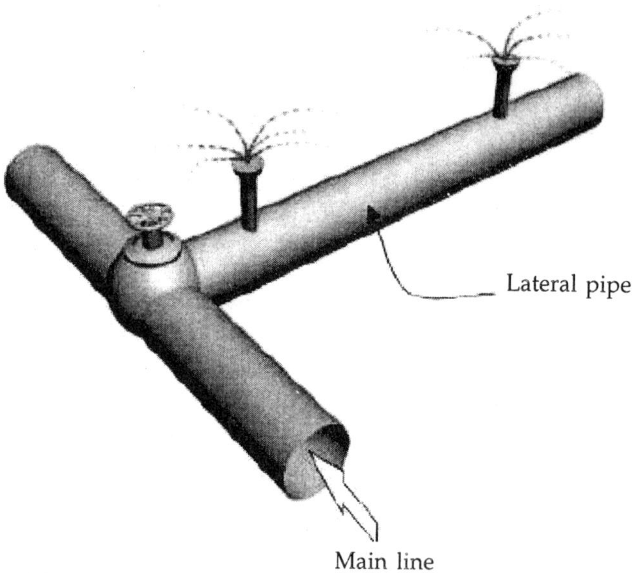

Figure 4. Rotating head system of sprinkler irrigation

Although no statistics are available on the total area under sprinkler irrigation, more than 200000 sprinkler sets were sold between 1985 and 1996 according to the National Committee on the use of plastics in agriculture.

The average growth rate of sprinkler irrigated area in India is about 25 percent. The cost of installation of sprinkler irrigation depends on a number of factors such as type of crop, the distance from water source.

Drip Irrigation System

Drip Irrigation system is sometimes called trickle irrigation and involves dripping water onto the soil at very low rates (2-20 litres per hour) from a system of small diameter plastic pipes filled with outlets called emitters or drippers.

Water is applied close to the plants so that only part of the soil in which the roots grow is wetted, unlike surface and sprinkler irrigation, which involves wetting the whole soil profile. With drip irrigation water, applications are more frequent than with other methods and this provides a very favourable high moisture level in the soil in which plants can flourish. Figure 4 shows a typical layout of the drip irrigation system.

A typical drip irrigation system consists of the following components:

— Pump unit

— Control Head

— Main and sub main lines

- Laterals

— Emitters and drippers

The drip irrigation system is particularly suited to areas where water quality is marginal, land is steeply sloping or undulating and of poor quality, where water or labour are expensive, or where high value crops require frequent water applications. It is more economical for orchard crops than for other crops and vegetables since in the orchards plants as well as rows are widely spaced. Drip irrigation limits the water supplied for consumptive use of plants. By maintaining a minimum soil moisture in the root zone, thereby maximizing the water saving.

A unique feature of drip irrigation is its excellent adaptability to saline water. Since the frequency of irrigation is quite high, the plant base always remains wet which keeps the salt concentration in the plant zone below the critical.

Irrigation efficiency of a drip irrigation system is more than 90 percent. Drip irrigation usage in India is expanding rapidly. There is even some Government subsidy to encourage its use. From about 1000 hectare in 1985, the area under drip irrigation increased to 70860 hectare in 1991, with the maximum developments taking place in the following states:

— Maharashtra (32924 hectare)

— Andhra Pradesh (11585 hectare)

— Karnataka (11412 hectare)

The drip irrigated crops are mainly used to irrigate orchards of which the following crops are important ones (according to a 1991 survey):

— Grapes (12000 hectare)

— Bananas (6500 hectare)

— Pomegranates (5440 hectare)

— Mangoes

Drip irrigation was also used to irrigate sugarcane (3900 hectare) and coconut (2600 hectare).

Important Terms

Dam: A dam is a hydraulic structure constructed across a river to store water on its upstream side. It is an impervious or fairly impervious barrier put across a natural stream so that a reservoir is formed.

Spillways and energy dissipators: Spillway is a channel that carries excess water over or around a dam or other obstruction. An energy dissipator is a device that is used to

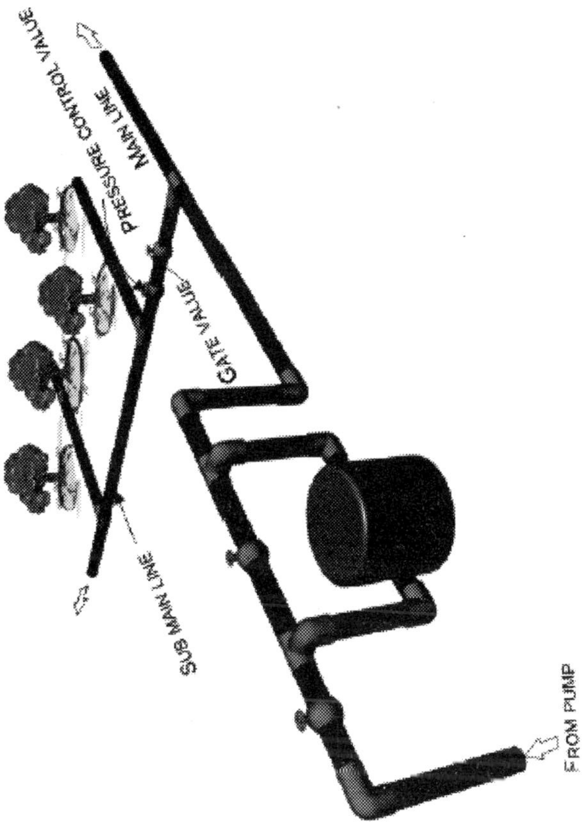

Figure 4. The tipical lay out of the drip irrigation system

convert concentrated storm water runoff to sheet flow and is constructed at the end of all storm sewers or channels that outfall into a buffer.

Sluice and Outlet: A sluice is an artificial channel for conducting water, with a valve or gate to regulate the flow. An outlet is a small structure which admits water from the distributing channel to a water course of field channel. Thus an outlet is a sort of head regulator for the field channel delivering water to the irrigation fields.

Barrage: An artificial obstruction placed in a river or water course to increase the depth of water. Canal Head Regulator: Any structure constructed to regulate the discharge, full supply level or velocity in a canal is known as a regulator work. This is necessary for the efficient working and safety of an irrigation channel. A canal head regulator regulates the supplies of the offtaking channel and the present channel respectively. The head regulator is provided at the head of the distributary and controls the supply entering the distributary.

River Training Works: Various measures adopted on a river to direct and guide the river flow, to train and regulate the river bed or to increase the low water depth are called River Training works. The purpose of the river training is to stabilise the channel along a certain alignment.

Cross regulator: A regulator provided on the main channel at the downstream of the offtake to head up the water level and to enable the off-taking channel to draw the required supply is called a Cross Regulator.

Pump Unit: This takes water from the source and provides the right pressure for delivering into the pipe system.

The control head: This consists of valves to control the discharge and pressure in the entire system. It may also have filters to clear the water. Common types of filters include screen filters and graded some filters which remove fine material suspended in the water. Some control head units contain a fertilizer or nutrient tank. These slowly add a measured dose of fertilizer into the water during irrigation.

Mainlines, sub mains and laterals: These components supply water from the control head into the fields. They are normally made from PVC or Polyethylene Glucose and should be buried below ground because they easily degrade when exposed to direct solar radiation.

Emitters or drippers: These are devices used to control the discharge water from the lateral to the plants. They are usually spaced more than 1 meter apart with one or more emitters used for a single plant such as a tree. For row crops more closely spaced emitters may be used to wet a strip of soil. The aspect of an emitter that should be kept in mind to decide its efficiency is that it should provide a specified constant discharge which doesn't vary much with pressure change and doesn't block easily.

References

Busman, J.D., *Optimizing control of surface irrigation using concurrent evaluation of infiltration*, PhD Dissertation, Agricultural and Irrigation Engineering, Utah State University, Logan, Utah. Unpublished document. 209p.

Doorenbos J. and W.O. Pruitt, *Crop water requirements*, Irrigation and Drainage Paper 24, Rome. 144p, 1977.

Garton, J.E., *Designing an automatic cut-back furrow irrigation system*, Oklahoma Agricultural Experiment Station, Bulletin B-651, Oklahoma State University, Stillwater, Oklahoma, 1966.

Gharbi, A., *Effect of flow fluctuations on free-draining and sloping furrow and border irrigation systems*, MS Thesis, Agricultural and Irrigation Engineering. Utah State University, Logan, Utah., Unpublished document, 123p, 1984.

Haise, H.R., Kruse, E.G., Payne, M.L,. and Duke, H.R., *Automation of surface irrigation: 15 years of USDA research and development at Fort Collins, Colorado*, USDA Production Research Report No. 179. US Government Printing Office, Washington DC, 1980.

Humpherys, A.S., *Surge flow surface irrigation: Section 3, Equipment*, Final report of Western Regional Project W-163, Utah Agricultural Experiment Station, Utah State University, Logan, Utah. 106p, 1987.

Ismail, S.M. and Westesen, G.L., *Surge flow border irrigation using an automatic gate*, Paper 84-2069 presented at the Winter Meeting of ASAE, Chicago, Illinois, 1984.

Jensen, M.E. (ed.), *Consumptive Use of Water and Irrigation Water Requirements,* American Society of Civil Engineers, New York. 215p, 1973.

Kemper, W.D., *et al., Cablegation system for irrigation: description, design, installation, and performance,* USDA-ARS Pub. 21, US Government Printing Office, Washington DC, 1985.

4

Modern Canal Operation Techniques

A major part of the 250 million ha irrigated worldwide is served by surface canal systems. In many cases, their performance is low to mediocre. There is a critical need for improvements in:

- — water resources management;
- — the service to irrigated agriculture;
- — the cost-effectiveness of infrastructure management.

Managing canal irrigation systems to achieve efficiency, equity and sustainability is a difficult task. Participatory approaches and management transfer reforms have been promoted widely as part of the solution for more cost-effective and sustainable irrigation services. In recent years, large agency-managed systems have been turned over partially or completely to various types of management bodies, which have had to struggle to improve service to users.

A vision of water services and modernisation plan for canal operation:

- — *The service to users:* This is the main purpose of the system management, and canal operation is the primary element in determining the service provided to end users. Service-oriented management (SOM) is the key for modern management; it does not necessarily imply a high level of service but the one that is best adapted to user demand. A clear vision of the water services should be the starting point from which others steps are carried out.

— *The re-engineering of management*: This includes reorganising the management setup and defining spatial units (partitioning management units) with the objective of favouring professionalism and cost-effective management.

— *Options for modernisation improvements*: The methodological development that can be used for developing a consistent strategy for improving canal operation and the project life cycle, in which managers and users need to engage progressively. It examines: analysis of the canal operation demands for the different units, the design of canal operation improvements, and a project to consolidate the improvements.

— A consolidated vision of the future of the irrigation system management and a plan for a progressive modernisation of irrigation management and canal operation.

Many canal irrigation systems perform well below their potential and improvements are needed urgently in water resources management, irrigated agriculture and asset management. In the last decades of the twentieth century, the emphasis was on performance outcomes and institutional reforms. This resulted in the management transfer of numerous irrigation systems and subsystems to water users associations (WUAs) and other farmer-oriented organisations. However, these new management bodies, formed as part of irrigation management reforms, often inherited dysfunctional infrastructure and severe financial limitations.

While documentation on the concepts and benchmarking of irrigation performance abounds, there are few manuals on canal operation techniques and ways to improve the water delivery service achieved by operators. Therefore, both public agencies and newly created water management bodies (e.g.

WUAs) are often ill-equipped to deal with the complexity of irrigation service delivery to users.

Importance of Separating Operation and Maintance

In 1976, Taylor and Wickham stated: "Separating operations and maintenance: although a certain degree of coordination between operations and maintenance is important to the smooth functioning of each, distinctions between the two must be made." This statement is still valid today. However, most of the time, an inadequate distinction is made between operation and maintenance (O&M) in terms of budget or responsibilities. Although they are quite different in nature, operation and maintenance have long been closely associated in irrigation management. While both apply to the physical infrastructure, operation differs fundamentally from maintenance. Operation is concerned with adjusting the setting of structures, whereas maintenance is about maintaining the capacity of the structures.

Therefore, it is important not to mix operation with maintenance. However, recognising and diagnosing trends or changes in the hydraulic properties of a canal form an intrinsic part of operations. The proper diagnosis should result in:

— an operational mitigation strategy; and

— hydraulic maintenance requirements/specifications to restore hydraulic and operational capacity.

There is a common misconception that canal operation is a well-understood and widely known technique, one that is well taught in engineering school and well mastered on the ground. Furthermore, there is the mistaken belief among many that the issues of poor irrigation performance are not related to engineering but more to do with the socioeconomic context. However, many surveys carried out by FAO show that canal operation is not well mastered and that it is very often the origin of the vicious cycle of poor service, poor fee recovery, leading to poor maintenance, and resulting in the physical deterioration of the irrigation infrastructure and services provided.

There is also a misunderstanding that the hydraulics and control techniques of canal systems are highly complex and always require the inputs of high-level experts, computers and a complex information system in order to achieve a reasonable level of performance. The truth lies somewhere in between.

Complexity of Canal Management

As a general trend, the complexity of irrigation management and canal operation has increased since the 1970s, mainly for three reasons:

— Service to users is more diversified. Improving the performance of irrigated agriculture requires more flexibility in water delivery for modern on-farm irrigation methods such as drip irrigation. Irrigation managers are increasingly confronted with a spatially diversified and dynamic service demand.

— Water management is more demanding. Increasing competition for water requires water management to be more effective and efficient. Complexities increase further where management evolves towards integrated water resources management (IWRM).

— Cost-effective management. Over time, it is becoming more difficult for governments to continue to subsidise irrigation management.

The period when direct and indirect inputs covered by government agencies were not really accounted for belongs to the past. Investments in irrigation infrastructure, state-owned or user-group-owned, need to be economically sustainable, and cost-effective management is now imperative. The complexity of operating an irrigation system depends on its physical nature and on the service expected. For open-channel delivery and distribution systems, which are the focus of the technical approach in MASSCOTE, the least complex types are those based on proportional division with few structures to be operated (also called "structured systems"), but where the service to the end user is minimum, inflexible and not differentiated.

Gated systems are more demanding in terms of operation but also provide a better range of service. One way to reduce the required manual efforts for delivering water is to introduce automation, which can be achieved through simple or very sophisticated techniques, and may or may not increase overall operational complexity. The operation of open-channel infrastructure is a complex task requiring numerous simultaneous or timely sequenced and coordinated actions along the canal network. It is demanding in terms of effort.

The nature of the efforts needed to operate an irrigation system is often, and it should be, adjusted according to the local technical and socioeconomic context. For example, in countries where labour costs are high and where most of the irrigation cost has to be borne by the users, many canal systems that were initially manually operated have been progressively automated to some degree. Automatic and self-acting structures performing with no or minimum direct human intervention should be based on sophisticated design techniques. However, the resulting structures can be simple.

In most countries, manual, labour-intensive operation of canal systems is still the prevailing method, but this manual operation can be improved and made more efficient and cost-effective. Thus, the choice for managers is not one of either "very expensive high technology" or "no change at all" but somewhere in between, and the necessity is to implement modernisation at an appropriate pace. Modernisation is a continuing process that requires step-by-step implementation and it must be driven by the demand and resources of users. Indeed, FAO (1997) has defined modernisation as: "a process of technical and managerial upgrading (as opposed to mere rehabilitation) of irrigation schemes with the objective to improve resource utilisation (labour, water, economics, environmental) and water delivery service to farms."

There is growing evidence that failure in specifically addressing canal operation and service-oriented management (SOM) in practical terms is a main reason for the lack of success in donor-funded modernisation programmes, management transfer and other irrigation sector reforms. The bottom line is that engineering aspects, and in particular the specific skills needed for effective canal operation, are prerequisites for successful and cost-effective irrigation management.

Service Oriented Management

The primary goal of the operation of a canal system is to convey and deliver irrigation water to users according to an agreed level of service that is well adapted to their requirements for water use and cropping systems. This approach is embedded in the concept of SOM, which substitutes previous more top-down and rigid approaches.

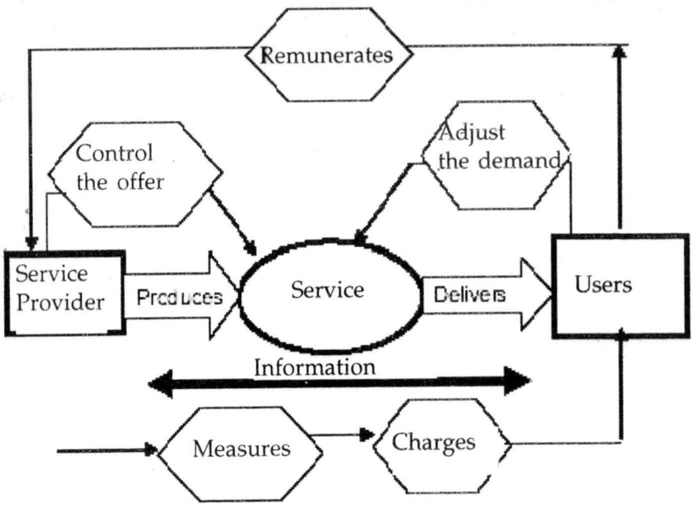

Figure 1. The service-oriented management approach

Service-oriented management can be modelled at the interface of agency–user (or supplier–receiver), as shown in Figure 1. In simple terms, the agency and the user first agree upon the specific details of the service of water (where, when, how, how much, etc.). The agency provides the service to the user, who in return remunerates the agency. It is generally considered that the effectiveness of a system in responding to user demands depends on its operational flexibility. Ideally, the users should be able to select and change the level of service corresponding to their demand, and the service provider should be able to control the delivered service to each user, and, if necessary, cut off the service in the event of nonpayment. This means that a key element in the concept of service is the information between the provider and the receivers, as well as among the receivers. Information is required in order to:

— predict the services that can be offered;

— assess the demand for services;

— correct the demand in real time during the season;

— adjust actual service to the demand;

— measure and charge for the services provided.

As regards to the service that should be remunerated, there are three basic flows in this SOM approach that must be considered:

— water;

— information; and

— money.

While canal operation is centred on water flows, it would be a mistake not to give full consideration to the other two elemental flows in developing new and/or improved canal operation strategies. The service of irrigation water also requires information flowing between the service provider and the users. Information is needed beforehand in order to agree upon a type of service, and then on a regular basis during the process of water delivery planning. All this depends much on the type of service. Where access to the service is free, the information flow needed for water delivery is minimal, if not nil. With an on-demand type service, the information must flow constantly in both directions. The request from the user goes up to the agency, then the service provider processes the demand, and a response goes down to the user.

MASSCOTE APPROACH

MASSCOTE seeks to generate a solution for irrigation management and operation that works better and that serves the users better. Canal operation is at the heart of the MASSCOTE approach for two main reasons:

- In the diagnosis phase: The critical examination of the canal state and the way it is operated yields significant physical evidence on the ground of what is really happening in terms of management organisation and service to users.
- In the development of the modernisation plan, canal operation is critical as the intervention aims to achieve the agreed upon and/or upgraded service. Many irrigation reforms have shown how important canal operation is the hard way, by neglecting it in the design.

Users are central to this SOM-based approach. The way the various steps of MASSCOTE are developed aims to generate solutions for service and operations on which the users will have to decide. Therefore, it is fair to say that canal operation is the focus of MASSCOTE, while its overall goal is modernisation of management and the users as central actors. Talking of modern irrigation management, it is always risky to bring forward a definition as there is then the possibility of not capturing all aspects of the problem, of being misunderstood, or of becoming rapidly obsolete or irrelevant in some context.

Canal operation is a complex set of tasks involving many critical activities that have to be carried out in a consistent and timely manner for good irrigation management. Among the numerous aspects of management, the following need to be considered:

- service to users;
- cost and resources dedicated for O&M;
- performance monitoring and evaluation (M&E);
- constraints on the timing and amount of water resources;

— physical constraints and opportunities relating to topography, geography, climate, etc.

There is no single answer as to how to integrate all the elements into an effective and sustainable framework for improving canal operation. However, the new MASSCOTE approach has been developed on the basis of extensive experience with irrigation modernisation programmes in Asia between 1998 and 2006.

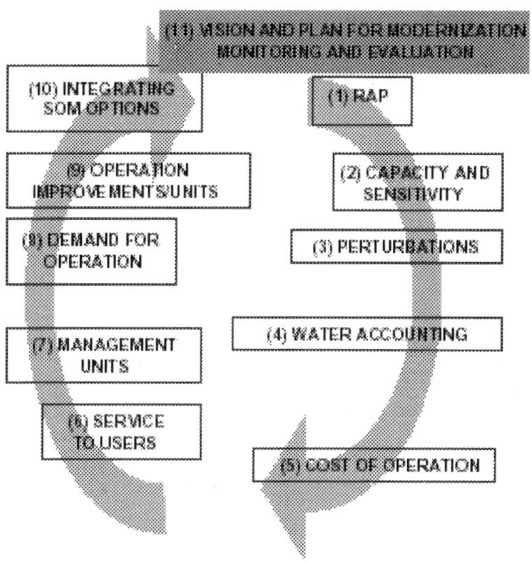

Figure 2. The steps in the MASSCOTE approach

MASSCOTE aims to organise the development of modernisation programmes through a step-by-step methodology:

— mapping various system characteristics;

— delimiting institutionally and spatially manageable subunits;

— defining the strategy for service and operation for each unit.

The first steps outlined in Figure 2 are to be conducted for the entire CA. The goal is to identify uniform managerial units for which specific options for canal operation can be designed and implemented.

Steps in the MASSCOTE Approach

Step 1: Mapping the performance

An initial rapid appraisal is the essential first step of the MASSCOTE approach. The RAP consists of a systematic set of procedures for diagnosing the bottlenecks of performance within an irrigation system. The RAP internal indicators assess quantitatively the internal processes, i.e. the inputs (resources used) and the outputs (services to downstream users), of an irrigation project. Internal indicators are related to operational procedures, management and institutional setup, hardware of the system, water delivery service, etc. They enable a comprehensive understanding of the processes that influence water delivery service and overall performance of a system.

Thus, they provide insight into what could or should be done in order to improve water delivery service and overall performance (the external indicators). The RAP external indicators compare input and output of an irrigation system in order to describe overall performance. These indicators are expressions of various forms of efficiency, e.g. water-use efficiency, crop yield, and budget. They do not provide any detail on what internal processes lead to these outputs and what should be done to improve the performance. However, they could be used for comparing the performance of different irrigation projects, nationally or internationally. Once these external indicators have been computed, they could be used as a benchmark for monitoring the impacts of modernisation on improvements in overall performance.

Step 2: System capacity and sensitivity

Mapping the system capacity and sensitivity deals with features of the physical infrastructure including the function of structures for conveyance, water level or flow control, measurement, and safety. Irrigation structures are intended to perform a particular function. How they are designed, installed, calibrated and maintained results in specific performance characteristics—some designs are better than others depending on the situation—and actual conditions may change with time owing to various phenomena, such as erosion, siltation and rusting. It is important to have a reasonable assessment of the existing status of the system in performing the basic functions.

Specifically, it is critical to identify any weak points, bottlenecks and/or areas with particular deficiencies. The mapping assessment of the flow capacity of infrastructure is necessary in order to compare with the design, but more importantly to ensure that the whole system is consistent with the operations plan to be developed. Any major structural deficiencies need to be addressed as part of the planning process of modernisation. Modernisation improvements cannot be carried out successfully without dealing with the impacts of severely degraded or dysfunctional infrastructure. Mapping the physical characteristics of the system is done in this step, and in particular the sensitivity of irrigation structures (offtakes and cross-regulators) is determined. Mapping of the sensitivity at key locations is crucial in managing perturbations.

The basic idea is to know where the sensitive offtakes and regulators are located, and which subsystems are propagating the perturbations and which ones are having to absorb them. Thus, in terms of mapping:

- mapping of structures: sensitive regulators and sensitive offtakes;
- mapping of subsystems: average characteristics per subsystem
- sensitive for flow control and water-level control.

This step gives rise to the following operational requirements and management options relating to sensitive structures/ subsystems:

- sensitive structures must be checked and operated more frequently;
- sensitive structures can be used to detect fluctuations (part of information management);
- sensitive subsystems can divert perturbations into subareas which are less vulnerable to lack or excess water.

Mapping perturbations

Perturbations of water variables (level and discharge) along an open-channel network are the norm not the exception. Despite being a target for canal operation, steady state along a canal is rarely found in practice. Thus, perturbation is a permanent feature of irrigation canals caused by upstream setting of structures, and compounded by intended or unpredicted changes in inflows/outflows at key nodes. Thus, if perturbations are unavoidable, then the only option for managers is to have a reliable knowledge of their origins, and to know how to detect and manage them. Managing a canal also deals with uncertainties and instabilities.

The types of perturbations that need to be mapped are:

- positive perturbations:
  - nature (inflow-outflow - internal),
  - magnitude (water-level fluctuation - relative discharge variation),
  - frequency;
- negative perturbations:
  - nature (inflow-outflow - internal),
  - magnitude (water-level fluctuation - relative discharge variation),
  - frequency.

With positive perturbations, the management options are:

- share the surplus proportionally among users;
- divert and store the surplus into storage capacity.

With negative perturbations, the management options are:

— compensate from storage;
— check for immediate correction;
— reduce delivery to some offtakes, with compensation later on.

Step 4: Water networks and water balances

In this step, the concept is to map the surface water network including irrigation and drainage layout, but also any natural channels if they interact or may interact in the future with the canal system and/or storage facilities. The objective is to know where and when all the inflow points to and outflow points from the service area occur in terms of flow rates, volumes and timing. This mapping includes all safety structures built to evacuate surplus water to the drainage network.

Managers must have accurate knowledge about all the paths of water (surface and groundwater)—where it is coming from and where is it flowing to, and in what volume. Knowing the water balance of the system is important not only for achieving high efficiencies but also for tackling environmental issues such as waterlogging and salinity buildup. It is also a good management tool for transparent water distribution within and among subareas of a system.

Step 5: Cost of O&M

In this step, mapping is done of the costs for current O&M. It also involves desegregating the elements entering into the cost and developing costing options for various levels of services with current techniques and with improved techniques. In order to produce the service that has been decided/agreed upon with users, managers need to mobilise a set of various resources or inputs, such as water, staff, energy, office, communication, and transport. All of these entail a cost. This step aims at clarifying the issue of inputs and costs for operation as part of the overall management activities and as fundamental elements of the modernisation process.

Investigating inputs and costs is important for:

— setting the service levels, in particular in exploring options for different types of services and associated costs;

— water pricing to users, in order to propose a set of charging procedures that takes into account the real cost of service production;

— improving performance and cost-effectiveness, by investigating technical options for maximising operational effectiveness (better allocation of existing resources, automation, etc.).

Step 6: Service to users

From the previous steps, a preliminary vision of the future of the scheme can be proposed for the future, from which initial features of the water services in the CA are derived:

— How many categories of service are considered, and how are these spatially distributed?

— How are the services evolving with time throughout the year?

— What is the service for crops with respect to the different seasons?

— What is the flexibility in defining the services with respect to the resources constraints?

— What are the features of allocation, scheduling and water deliveries that define the overall service?

Assessing all the different services provided to different users and their related costs are what need to be mapped in this step. Mapping of service is required for further analysis of modernisation opportunities and economic analyses to be done in later steps. This specific mapping exercise of services leads de facto to crafting a preliminary vision of the irrigation scheme which should be made explicit before carrying out the next steps.

Step 7: Management units

Canal irrigation systems serving large areas are usually divided into smaller manageable units called tracks, blocks or subsystems. In the past (and particularly for new systems), these management units have often been based on the hierarchy of the canal network (main, secondary, tertiary, etc). Today, with the increasing complexity of management and operation needed to provide higher levels of service, this partitioning might be less relevant than it was when the systems were originally constructed. There are more relevant operational criteria on which subunits should be based, for example:

— participatory management;

— spatial variation of water services;

— conjunctive water management;

— multiple users of water;

— drainage conditions.

Subunits of operation/management should define an area for which a certain level of service is agreed upon and provided, and for which the water balance is to be managed as a single unit. A workable compromise has to be found between the physical/ hydraulic system and the institutional/managerial resources in each subunit. The grounds on which subunits should be based are multiple. However, the setting up of too many units should be avoided, keeping in mind the baseline costs associated with the management of individual units.

Step 8: Demand for operation

This step involves assessing the resources, opportunity and demand for improved canal operation. It entails a spatial analysis of the entire service area, with preliminary identification of subsystem units (management, service, O&M, etc.). Assessing the requirements for canal operation needs to be done alongside and in combination with the definition of the service by users and stakeholders. However, canal operation requirements cannot be derived only from service demands.

The system presents opportunities and constraints that set the boundaries for possible modes of operation. In short, the requirements for operation will depend on three domains:

- the service will specify the targets;
- the perturbation will specify the constraints in which the system operates; and
- the sensitivity will specify how fast the system reacts to changes and produces changes.

The rationale is straightforward: the higher the sensitivity, perturbations and service demand, the higher the demand for canal operation. This can be expressed in the relationship: demand for operation = service × perturbation × sensitivity.

Step 9: Canal operation improvements / units

This step entails identifying options for improvements to canal operations. Improvements should aim at specific objectives such as:

- improving water delivery services to agriculture users;
- optimising the cost of operation;
- maximising the conjunctive use of water;
- integrating the multiple uses of water (IWRM).

It is necessary to develop modernisation improvement options for each subunit based on:

- water management;
- water control; and
- canal operation (service and cost-effectiveness).

The improvements are to be sought through one or a combination of the following options:

- allocating existing resources and inputs in a more cost-effective and responsive way;
- optimising the organisation and the operational modes;
- changing the operational strategy;
- investing in improved techniques and infrastructure.

For water management, the improvements aim to increase productivity and/or storage by:

- minimising losses;

— maximising harvest; and
— re-regulating storage.

For water control, the improvements concern the hydraulic configuration of the operations. This entails a sequence of:

— fine-tuning the hydraulic heads of canal structures in relation to each other;
— creating a specific hydraulic property of the canal (section) so that it performs as intended; and
— choosing the option that will minimise manual operational interventions/regulations for a specific period.

Step 10: Integrating service-oriented management options

Improvement options for the subunits are finalised together with the associated costs for every option. These are then aggregated for the entire command area in line with the improvement option at the main canal level. A modernisation strategy is laid out with objectives and proposed achievements/improvements.

Step 11: Plan for modernisation and M&E

The carrying out of the previous steps with some reiterative cycle is the process by which, progressively, a vision of the future for the irrigation scheme is crafted and consolidated. This vision must then be converted into a plan that should aim at implementing the vision. Modernisation improvements must be implemented in order to keep expectations and potential achievements at a realistic and practical level. A decision about the options to pursue is taken through extensive participation of the users. The solutions that are easiest and most cost-effective to implement are to be selected to start the process of modernisation.

Monitoring and evaluation of the improved operations are necessary in order to ensure that achievements are maintained, and to provide a basis for comparison of the situation before and after the improvements.

Features of MASSCOTE

There are four important features to bear in mind about MASSCOTE. The first is the embedded nature of the RAP and MASSCOTE within a modernisation project. (Figure 3).

The second feature concerns the different time frames of the interventions:

— RAP = week;

— MASSCOTE = month;

— modernisation project = year.

The third feature concerns the revolving nature of MASSCOTE. This might imply iterative circles before reaching a consolidated stage of analysis and project—several rounds of MASSCOTE at given levels before integrating at the upper level and going back at lower level.

Figure 3. Embedded nature of the RAP and MASSCOTE

The fourth feature is that a major entry point of the MASSCOTE methodology is canal operation, for diagnosis and for designing improvements. However, the overall objective in carrying out a MASSCOTE exercise is modernisation of management. Canal operation is a critical entry point because:

— it is the activity that puts management decisions into tangible outputs; and

— it is there that the current management performance is sanctioned and expressed in the most obvious manner (its symptoms).

Field survey along a canal system is the most effective and reliable way of identifying management problems. MASSCOTE evolves from canal operation to management options (institutional partitioning, organisation, and SOM).

RAP and Benchmarking

The MASSCOTE approach needs to be seen in the context of other irrigation management and modernisation tools and methodologies that have been developed in the last decade, in particular, the RAP and benchmarking. These approaches are developed in the same three-dimensional space of impact (external indicators), process (internal indictors) and solution (option for improvements). The focus might be different, and some approaches are more inclusive.

Benchmarking allows monitoring and checking of the performance of the management compared with other similar systems elsewhere, or after having introduced some improvements in the techniques and procedures. It is an essential component of a modernisation project development. The MASSCOTE approach adds value to benchmarking and the RAP by focusing on the development of solutions that are derived from a thorough diagnosis of the impacts and processes that the other two tools provide. Therefore, it is logical that the first step in the MASSCOTE approach is the RAP.

Irrigation Canal Operation

Irrigation canal operation depends on various factors related to the types of:

- systems (gated and ungated);
- control (mainly upstream control and downstream control);
- operation (manual, motorised and automatic);
- service delivered to users (rotation, arranged, free access, etc.). Operating an irrigation system consists of carrying out a specific set of actions at the control and measurement structures (hardware) of an irrigation infrastructure network in order to:
- convey, deliver and monitor water to meet a pre-defined irrigation service to end users/clients,

according to the schedule and the allocation agreed upon;
- ensure efficient water management within the gross command area;
- maintain the infrastructure/hardware.

Thus, operation is not limited only to physical interventions at major structures. It also includes:
- information collection from users for water orders and water charges;
- regular observations on the status of the system;
- decision-making procedures with user participation;
- M&E of the effectiveness of implementation.

Purposes of Operation

The purpose of canal operation is manyfold:
- Scheduled operation for planned setting changes according to updated water distribution plans. Actions at this level aim to provide the targeted water delivery service. This mode of operation is also called predictive operation.
- Routine operation to deal with stabilising perturbations by making changes in the settings of control structures for water supply and delivery. The perturbations are caused by illegal/unforeseen interventions, or difficulties in predicting natural causes (floods, winds, rainfall, and increased return flows). Actions at this level are undertaken in order to react to unplanned changes, with the overall objective of maintaining the quality of service as well as ensuring the safety of the system. This mode is also called reactive operation.
- Emergency operations. When unexpected surplus water in the canal system creates the risk of breaches, emergency spill structures have to be activated (where they are not automatic).
- M&E of the process at regular intervals is necessary for sound decision-making by the operators, and it is

essential for evaluating the service to the users. Therefore, M&E deals with the status of the system structures (intended vs actual) and flows at key points, as well as the service provided to users. Actions target frequent monitoring of the internal physical variables (water levels, discharges, and gate settings) and the service (deliveries to intermediate and/or end users).

In the category of scheduled operation, different types of interventions can be distinguished:

- — restart of irrigation deliveries (filling the canals at the start of the season or between rotation cycles);
- — regular water distribution changes;
- — de-watering at the end of the season (canal closure). For each type of canal operation, a specific procedure (or set of procedures) needs to be established as part of an operations plan.

In practice, each category of operation aims to achieve a specific objective. For example, the targeted service to users defines a water distribution plan (WDP), which basically specifies the flow rate at each key location of the system as a function of time.

Functions of Canal Structures

Operation is a set of actions at irrigation structures to perform specific functions. A hydraulic infrastructure network is a set of interconnected structures, each one ensuring one or several specific functions. The structures of a network serve the following functions:

- — storage,
- — conveyance,
- — diversion,
- — distribution,
- — control,
- — measurement,

safety,

— transmission.

The storage function

The storage function consists of storing excess water at a given point in time and space (runoff, and discharge in rivers or canals) in order to deliver it at a more convenient time and place according to users' requirements. The lag time between storage and distribution may have different time steps, ranging from a few hours (night/day) up to some years for reservoirs that ensure several years of regulation.

The storage function is often ensured by surface reservoirs behind a dam. A distinction can be made between storage reservoirs situated upstream of the service area and inline or intermediate regulating reservoirs. Proper use of the storage function results from the coordinated release of water in relation to the capacities of canal system. Finally, the storage function today cannot ignore the utmost importance and great vulnerability of groundwater. Aquifers sometimes represent an important and usable storage but may be equally limited in recharge. Today, the protection and management of underground aquifers (control of withdrawals, and recharge of the groundwater) are a critical part of the issues facing water resources managers.

Conveyance function

In most irrigation systems worldwide, conveyance is made through open channels. However, there are also buried pressurised pipe networks, and buried gravity networks (as in the traditional systems of piedmont groundwater abstraction, such as Khetarras in northern Africa). Natural systems (rivers) are also used to convey water between storage and the place where it is diverted to be distributed through the irrigation network.

Diversion function

This is the function by which irrigation water is diverted to be conveyed to the area where it will be used (irrigation scheme or subscheme). Diversion works are installed either on rivers or on large conveyance canals. Where the withdrawal is made on a dam, it is typically called an "offtake" structure. On rivers, the structure is often called a "diversion dam", but it usually has a very limited storage function; its essential function is to raise the natural water depth in order to supply water to the intake canal by gravity.

Distribution function

Distribution consists of delivering the required discharge to key points in the network (head of secondary, tertiary and quaternary canals). This function is typically accomplished through gated structures that divert a regulated discharge from one canal level to the next lower level.

Division function (proportional)

In proportional irrigation systems, the flow is divided proportionally at key points in order to allow a preset share of the available water to be distributed to downstream branches.

Control function

In order to ensure the good operation of a conveyance and distribution network, some intermediate variables need to be controlled. For example, on a pressurised network, the pressure is controlled at different points. In the case of an open structure, the water depth is controlled in the canals, in particular, close to the offtakes. Control structures are equally called regulators, cross-regulators, level regulators and check structures.

Safety function

The infrastructure in a canal system branches as it goes downstream and the conveyance capacity of individual structures is reduced. Owing to the nature of unsteady flow, it is necessary to ensure the safe disposal of spill water. In an upstream control canal system, such an overload can exceed the capacity of the conveyance structures. It is then a matter of performing, at some critical points, the disposal of all the additional discharge in order to prevent any damage to the canal and the areas it passes through (risk of breach in the canal, and flood hazard for riparian areas).

Measurement function

Management of canal systems entails regular decision-making with respect to the known status of the system. Therefore, it is necessary to obtain information on the state of the system in order to organise a proper response. Thus, monitoring at key points in the system through appropriately designed and situated measurement structures is essential to the manager. These structures have to quantify accurately relevant parameters that are important for management.

Information transmission function

This function aims to ensure that information collected in the field is available in real-time or near real-time at the decision-making centres. This function is being performed increasingly by wireless communication devices. Supervisory control and data acquisition (SCADA) is the system often referred to in relation to information and control along an irrigation system.

Information management function

Although not a part of physical canal system compiling, processing, displaying and archiving are the basic functions of information management.

Types of Canal Systems

Irrigation systems are composed of numerous reaches—conveying flows and nodes that are division or diversion points. A node (also called a bifurcation point) is a particular point where:

- the flow in a canal is subdivided into two or more flows according to a preset pattern or to a specific, controllable target;
- the error/deviation from targets is also subdivided into the different dependent canals.

Thus, a node is defined with the specific flow targets of each branch, but also by the way deviations are shared. The node can be proportional, overproportional or underproportional. There are two basic categories of nodes: gated or ungated, and this corresponds to the two main types of systems: gated, and ungated. The latter are often based on a fixed proportional division of the inflow, typically called a proportional system. The former are equipped with adjustable gates that are used to adjust the outflow from zero to the maximum value.

Controlled Water Variable

Operation consists of manipulating gates and structures in order to produce the agreed service and deliver it to the users. There are various types of control logic for canal operation, depending on several features. One important aspect is to define the control variable. Discharge control is the most common control procedure whereby inflows from the main inlets are adjusted to match the discharge demand along the infrastructure, or the deliveries are adjusted to match the net availability (inflows minus losses). This technique of "discharge control" goes together with the control of water depth in the canal to ensure a steady head at the outlets.

Other systems are designed and operated to control volume in canal reaches. This technique requires the availability of storage, either inline storage capacity in the canal itself or in intermediate reservoirs. Available storage is dependent on variations in the water depth in the system. Therefore, offtake discharge should be somewhat independent of upstream water level, i.e. outlet structures should have a low sensitivity to the changes in water level in the parent canal.

Type of Control

Most gravity irrigation systems are based on upstream water-level control. With this technique, cross-regulators in the canal have to be adjusted on a timely basis in order to maintain the water level immediately upstream owing to variations that arise from changes at the headworks, considering the time lag for water transfer and changes to the flow diverted by upstream canals or turnouts or entering the canal. The objective is to maintain the water level upstream of each cross-regulator in order to control the backwater profile in the upstream reach. The backwater profile determines the head at offtakes in the upstream reach.

The alternative technique, i.e. downstream control, has attracted the attention of engineers and irrigation managers mainly because of the potential advantage of responding automatically to varying downstream demands from users. However, the technique is expensive as it usually requires horizontal canal banks and automated control structures.

Types of Operation

Manually operated systems

For a manually operated gated system, irrigation staff have to manipulate every offtake and control regulator when a change in the flow regime is scheduled or occurs because of an unscheduled perturbation. This task has to be carried out at least once per day. The difficulty in operating these systems results from the numerous structures to be adjusted when the flow regime is changing. This large number of structures implies the mobilisation of correspondingly large amounts of resources (human and/or transport) for adjusting and monitoring control settings. The greater is the density and sensitivity of structures, the greater is the difficulty of the control task resulting from unsteady flow conditions.

Ungated systems are easier to operate from the standpoint of the system operators as they do not require numerous and frequent interventions for regular operation. In the commonly known systems originally developed in India, Pakistan and Nepal, typically termed "structured systems", water delivery is organised around releases of constant discharge with a varied frequency. Distribution is proportional below the structured point, and structures are permanently fixed at the construction stage (no adjustable parts).

The savings in resources for manipulation of structures can be large. These systems were developed mainly for conservative irrigation and famine protection, with the goal of serving an average of one-third of the water needs for the entire CA. At the time of their construction, they were modern in the sense that they were responding to the urgent needs and matching the resources of their time. Today, with increasing demand for crop diversification and with rapid growth in cropping intensity, they are often no longer able to satisfy user demand.

Automatic semiautomatic gated systems

Automated systems are equipped with structures that control the water levels in canals over a full range of discharge. These structures may be either downstream or upstream control devices.

The control of water level is achieved by mechanical movements of regulator gates, slide gates, radial gates, and flap gates. Automated systems differ by the way gates are operated. Generally speaking, there are:

— energy-driven gated systems; and
— gates driven by hydraulic forces without an external source of energy or human intervention.

In many Mediterranean countries, several modernised systems are equipped with hydraulic driven gates. In the United States of America, the gates are more often motorised, with a local programmer controlling the water level. The hydraulic-driven gates include AMIL, AVIS/AVIO, DACL and Danaidean gates. Variations in water level must still be expected to occur at locations remote from the control regulator. Hence, hydraulically automatic cross-regulator structures are frequently associated with constant discharge distributors, such as baffles, in order to enhance overall performance.

Some cross-structures can ensure good control of the water level without gates. They use a simple long-crested weir (LCW), which minimises drastically the variation in water level upstream caused by discharge changes to the extent that this variation is acceptable for the nearby offtakes.

One category of LCW is the well-known duckbill weir (DBW). In these systems, the water level upstream of the LCW structures is controlled when the canal flow varies. Therefore, the discharge variation through the nearby offtake is minimised by selection of low-sensitivity offtake structures.

Simple pipes in the bottom bed between the parent canal and the dependent canal are also simple ungated offtaking structures, whose performance depends on the head exercised.

Structures of Gated Systems

The most common structures in gated systems are:

— offtakes (diverting structures), to control water diversion at a given point; and

— regulators (water-level control structure), to minimise water-level fluctuations at a given point.

If the offtake is not sensitive to water depth variation in the parent canal, then there is no need to install a cross-regulator. This is the case for some specific structures such as the baffles, but also for some orifice-type offtake when they are fed with sufficient head (H), say 1 m or more (meaning they are "low sensitive").

Where offtakes are sensitive to water-level fluctuations, it is often necessary to control the level at this node by installing a control structure.

The adjustment properties of irrigation structures are:

— the freedom and precision that can be exerted in the adjustment of the structure;

— the effort required for manipulation and control; and

— the hydraulic stability based on the sensitivity of the structure. These properties lead to the identification of the criteria for operation.

The properties freedom of adjustment and precision of control can be analysed through the classification of structures as proposed by Horst:

- — Fixed: no adjustment is possible, e.g. weirs, orifices and dividers.
- — Open/closed: generally gates for minor canals, either fully open or closed.
- — Step-by-step: regulation by steps, modules or stoplogs.
- — Gradual adjustment: gated orifices, and movable weirs.
- — Automatic: hydraulically adjusted gates.

For fixed structures, freedom of adjustment is nil as output is imposed directly by ongoing discharge (input), and precision is meaningless. For open/closed structures, freedom and precision are not relevant. For step-bystep adjustment, freedom and precision are limited by the number of discrete steps in the adjustment between zero and full capacity.

For gradually adjustable structures, the degree of freedom is intrinsically high in that it is generally possible to choose any setting between zero and the maximum value. Precision will depend on the increment of the mechanical adjustment. For hydraulically automatic structures (self-acting), flow conditions are the governing factors.

In general, these structures cannot be adjusted in normal use and, therefore, the degree of freedom is zero. However, the operational objective is to maintain constant output, and precision is determined by the range of variation in output resulting from variations in input. Finally, for all types of structures, it is necessary to distinguish between manually, hydraulically, and motorised control structures.

Operation of Irrigation Structures

There are two critical steps in organising the operation of a canal:

- — defining the specification of operation for each structure (considered as independent);

— defining the sequencing of interventions: operation plan for scheduled change and foı routine interventions.

Operating a structure means a cycle of various activities:

— decision to operate;

— modalities of operation;

— intervention in the structure; and

— monitoring of the structure, which can then again trigger a decision to operate, etc.

The specific function of the structure can be to control the diversion flow, regulate a target water level, measure key variables, or record information. Different types of structures are used to perform these different tasks. For each type of structure, managers must define clear targets to be achieved and establish clear sets of instructions for operators on how to proceed.

Operating a delivery point (offtake) means achieving a time-bounded change in the discharge at this point. Where it is a single end-user outlet, it can be on and off, with or without the possibility for adjusting discharge. Where it is an intermediate node serving a large group of users, it can entail adjustments to allow a range of flows.

Operating a delivery point entails a set of physical interventions that are:

— manipulating the structure: opening and closing of the gate;

— adjusting for the targeted discharge: setting the gate opening;

— checking and reacting.

Manual operation implies that an operator must be present at the structure in order to manipulate the gate (open and close) according to the distribution plan and also in order to perform routine operation. Thus, "operation" mobilises various types of resources: staff, transport, communication, capacity and instructions. Given the numerous structures along a canal system, the physical operation of one single structure has to be put into the context of:

- The decision-making at the management level. Specific schedules and targets have to be decided according to the water distribution plans and water balance of all inflows and outflows (canal and management losses).
- The infrastructure network, where interactions among structures, time lags between action and effects have to be taken into consideration in order to minimise the requirements for interventions (or stated another way, to maximise their effectiveness).
- The coordination of resources allocated/available to operate the system. A singlestructure operation is simple to perform where means are sufficient, e.g. staff can be deployed at each structure or group of nearby structures. However, complexity arises where there are many structures within the CA. This requires a well structured organisation to coordinate and optimise operations while minimising O&M costs.

In upstream-control systems, the objective is to control the water depth upstream of the regulators within a specified variation (tolerance) around the target. This target has usually been set to allow offtakes under the influence of the regulator to be fed properly. Cross-regulators can be fixed (LCW), automatic (AMIL gate) or adjustable, consisting of one or more gates. Apart from a few exceptions on modern systems, cross-regulators are often equipped with undershot gates. A significant improvement is obtained with undershot gated regulators where they are equipped with dual side weirs. In this case, the objective for operation is to keep the water surface slightly overtopping at the spill level of the side weirs.

The gates of adjustable regulators must be operated with specific rules (reaction to a measured deviation of water depth, to the pace of changes, etc.) in order to enable good control of water depth without generating too many oscillations of the water profile along the canal. In a manually operated system, specific rules, although much simpler, must also be worked out. The operation of a regulator consists of mainly two elements: the timing (when to operate); and the mode of adjustment (how to adjust).

In manual operation, it is common, for routine operations that the correction applied by the operator to the gate setting is proportional to the observed deviation of water level from the target, which corresponds to full supply depth (FSD). In describing an operational procedure, it is necessary to distinguish between:

— scheduled changes in flow rates or predictive operation (which require direct adjustment of the regulator gate settings to allow the expected discharge at this point after the water surface profile has stabilised); and

— routine operation or reactive operation.

Operations for emergencies and M&E are quite different by nature, and they are not considered here. With a frequently operated system (often automated), there is no need to distinguish between scheduled and routine operation; each cross-regulator routine operation; each cross-regulator is operated according to the measured variations by sensors. With manual operation, it is important to make the distinction between scheduled and unscheduled operations.

Operations at System Level

In large canal systems, structures are highly interactive. Operations are not merely the addition of independent actions. Rather, they must be a coordinated set of actions aimed at maximising the service to users and minimising losses. With medium to large systems, the implementation of canal operations is not done solely by one person or one group but split into multiple operational units. These units are defined in several ways:

— partitioning into clear-cut separate water management units;

— administrative district/sectors;

— groups of major canal structures that can be handled by one operator.

Clear-cut defined separate water management units are intended to allow for independent water management and canal operations in a defined zone. For the latter two, management and operation are more dependent on what is happening upstream.

Distribution plan

Before the operation of a scheduled delivery on a weekly/ daily basis, planning has to be done based on demand analysis, possibly emanating directly from the users (water ordering) and some aggregation, in order to ensure the balance of available water supplies. An operation plan is also necessary in order to ensure proper allocation of means (transport, staff communications, etc.) and hydraulic smoothness of the planned change.

Scheduling of deliveries and flows at main points and nodes The scheduling of deliveries and flows requires an operation plan (a consistent sequence of interventions on the structures). A typical motivation for scheduled adjustments to the cross-regulator structures along a canal is when there is a change in the distribution pattern (e.g. every week or fortnight). For example, this happens when the rotation of water deliveries is changed, implying an increase in flow at some delivery points, and elsewhere a decrease or a cutoff (if on rotation). The whole balance of water flows has to be moved from one stage to another. This implies numerous changes on control and delivery structures. Such changes need to be organised in a coordinated and effective way.

From Water Distribution to Operation Plan

There are many ways of operating a system depending on the water management constraints and opportunities, the techniques in use, the physical conditions of the system, etc. However, all gated upstream controlled systems follow the same basic steps from the comparison between the demand and the capacity, down to the operation plan via the water distribution plan (WDP), as illustrated in Figure 4.

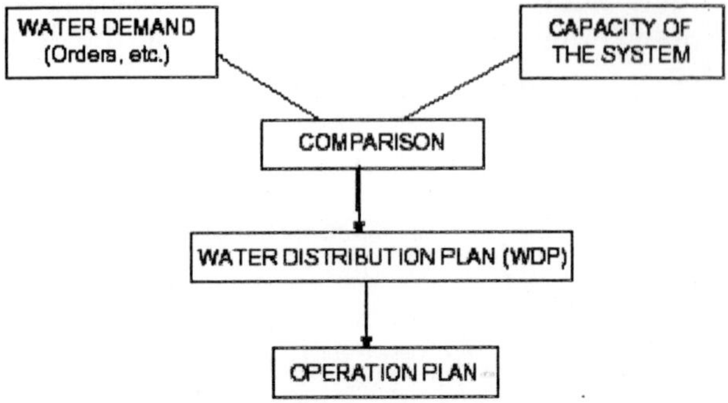

Figure 4. From water demand to distribution plan down to operation plan

The WDP is the first step in developing a canal operation plan. It is constructed around matching the users' requests with the constraints of the available water resources, as well as with the capacity of the infrastructure for conveyance and distribution:

- — Collection of water orders from users and demand analysis for water services.
- — WDP (per day, per week and or longer [ten days or monthly]): a time-based and location-based allocation of water flows and volumes within the service area, and throughout the canal system, considering constraints on water availability, and physical constraints of conveyance.

The operation plan aims to implement the WDP while considering three important features:

- — the scheduling of water deliveries at delivery points according to the WDP;
- — the necessity of dealing with errors and uncertainties;
- — accommodating unscheduled changes.

As a result, an operation plan must have a consistent system-wide procedure/ organisation/sequence in order to:

- — implement scheduled changes;

— deal with uncertainties;
— have local instructions that can take care of unscheduled changes.

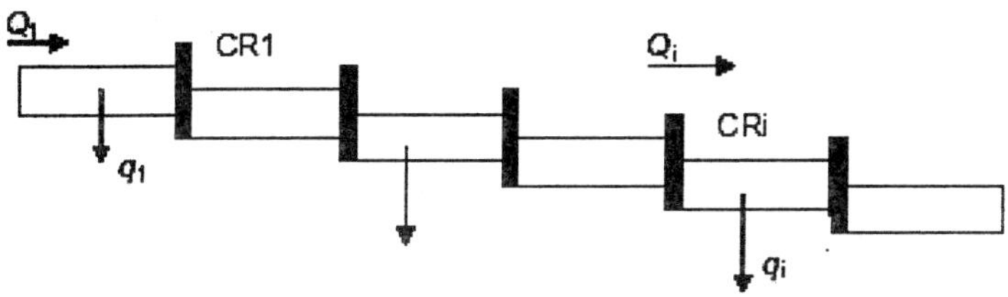

Figure 5. Sketch of a canal as a set of pools

Figure 5 shows a sketch of a canal as a set of pools. With this as an example, the following questions may be posed:

— How to organise the sequence of operation at cross-regulators to allow a change in withdrawal in the reach "i", for example by opening a new offtake discharge from 0 to qi at a given time ti?
— When should operators change the main discharge at the headworks?
— What is the sequence of operations at the cross-regulators between the headworks and the reach "i" that should be implemented in order to put the new distribution pattern in place?

Scheduled/Predictive Operations

For a simple cross-regulator along a canal, the literature mentions several procedures for scheduled or predictive operations. The main ones are:

— sequential downward, which includes: time-lag operation (TLO), or any variation of TLO, such as proportional to time lag (PTL);
— sequential bottom-up (SBU);
— simultaneous operation (SO).

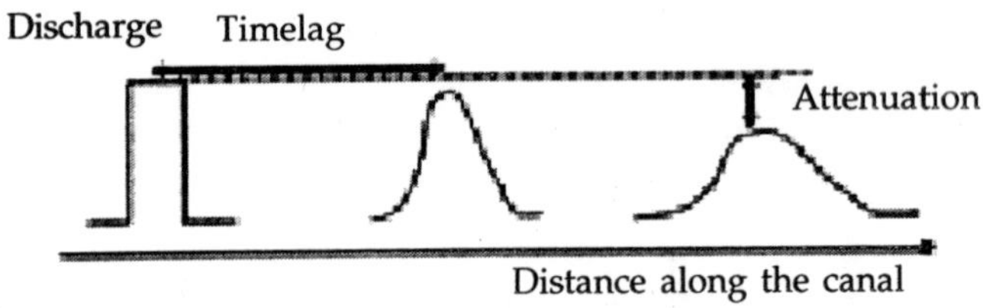

Figure 6. Downward wave propagation along a canal

Sequential TLO

The sequential downward regulator operation is particularly compatible with manual operations as the operator can adjust the gates sequentially while travelling down the canal. The sequential TLO requires gate operators to adjust gate settings as the transient wave front arrives at the cross-regulator in response to upstream operations. With this technique, the anticipation of the passage of the transient wave is zero. Changes in withdrawals must wait for the passage of the wave. The transit times of changes can be relatively long in canals. It is not rare that a change in supply to a long canal takes more than 24 hours to become apparent at the downstream end of the network. In order to operate structures and meet demand on time, it is crucial for managers to know how the waves are propagated through the system (Figure 6).

Transfer time from the main reservoir to any point along the infrastructure can be estimated from past experience or from evaluation using a non-permanent model. Detailed knowledge of the transit time along a canal system can be translated into a management strategy of structures and, in particular, it serves to prevent established management rules from giving rise to amplification of perturbations along the canal (oscillations).

The difficulty with this method arises when the time lag extends beyond 12 hours, which is the case for most medium to large systems. This means that, somewhere, operations may have to be carried out at night—which may sometimes be socially difficult and not easy. When the time lag to reach the tail-end exceeds one day, then this method creates large delays in enabling delivery changes, which may or may not be compatible with the water schedule. Where these delays are not acceptable, the managers should proceed with anticipation of the deliveries by issuing in advance the flow changes required for the tail-end; hence, the time-lag change is no longer applicable.

Bottom-up Per Block

The bottom-up operation consists of implementing gate adjustments starting from the tail-end of the system. In practice, each gate operator is responsible for the operation of several cross-regulators (termed here as one block). Therefore, the operators start adjustments at the same time at the most downstream of the regulators under their individual control. After setting the required gate position, the operators move to the next regulator upstream. In general, the delay between operations of successive regulators is about 60 minutes. Finally, the main intake regulator is adjusted, and the change in the supply propagates through the system. Anticipation of the wave is maximum in bottom-up operation.

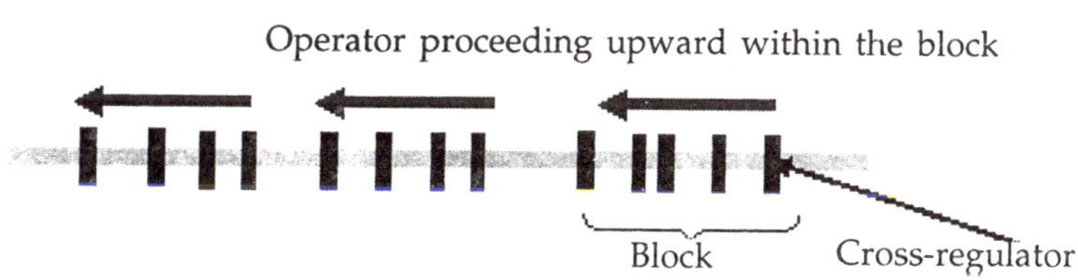

Figure 7. Illustration of bottom-up operation per block

Simultaneous Operation (SO)

Simultaneous operation (SO) requires that all structures be adjusted at the same time. This enables a new steady state to be established rapidly along the canal. When operated, regulators generate both positive and negative waves in the adjoining reaches. These waves cancel each other at the pivot point of the pool and establish a new steady profile. This is possible only where an operator is available at every structure. In practice, operators have to move from one regulator to the next.

Anticipation of the transient wave is intermediate between time-lag and bottom-up operations. Anticipated time-lag inputs with simultaneous operation When a delivery change has to take place at the same time of the week from a long canal, then a TLO cannot be applied as described earlier. The time lag has to be considered by anticipation. Incremental anticipated inputs changes from the main supply can set the system to the right status with the right flows at the time of the change. For example, an increment increase of 10 m^3 is made 12 hours in advance at the main supply if the time lag to reach the point of this particular delivery increase is about 12 hours. While the incremental inflow changes (waves) are passing through the canal, upstream regulators are operated on a routine basis.

Proportional to time-lag (PTL) operations are a compromise between TLO and SO. Gates are operated at a specified proportion of the time lag (between 0 and 1). The degree of anticipation is variable. Implementation of PTL operation requires operators to have a rough estimate of the usual time lag. This can be obtained experimentally by observing the propagation of a flow change along the canal. These estimates can thereafter be used to identify approximate values for the PTL at each cross-regulator of the canal.

Routine Operations

Routine operations are carried out at cross-regulators only and they occur at a fixed frequency (FF). For example, in Sri Lanka, the frequency of operation is often generally twice per day, one operation taking place between 7 and 9 a.m. and the other between 4 and 6 p.m. This pattern corresponds to a nominal 12-hour frequency of operation. Exchanges of operational information between gate operators and the system manager are limited to one exchange per day, usually in the morning.

With the FF procedure, no specific operations are identified for response to unscheduled flow changes; routine adjustments at a frequency of 12 hours are considered sufficient response. For instance under the usual mode of operation—with target set at full supply level (FSL)—no attempt is made to manage positive flow changes, for example, by storing additional flow volumes either in the canal section or in inline reservoirs. In that case the basic management objectives are to minimise the impact of flow changes on deliveries in progress and to dissipate peak flows without structural damage to the canal.

Emergency Operations

The aim of emergency operations is to prevent serious failures in the canal system caused by unexpected flooding, structural failures, etc. They do so by channelling or storing water surplus in natural streams and storage basins.

Monitoring and Evaluation

Monitoring and evaluation are required in order to enable sound decision-making for operations, and are essential for evaluating the actual service provided to the users. Therefore, M&E targets the status of the system structures and flows as well as the service to users. Actions here target monitoring of the internal physical variables (water levels, discharges, and gate settings) and the service (deliveries to intermediate and/ or end users).

Performance analysis is an intrinsic part of management. It is needed in order to target and monitor actual achievements in operation. Performance should be looked at from three perspectives:

- the service to the users;
- the efficiency in managing the resources; and
- the cost of managing the infrastructure.

Operation has its own, very specific, information requirements (collection, transmission and processing) and, thus, operation plans have to include specific "operational information management systems".

Refilling Canals at the Start of the Irrigation Season

When operations start at the beginning of the season, the system must be cleaned and all accumulated trash removed. This is particularly important in systems in urban areas. Some pre-cleaning must be carried out in order to remove most of the buildup. However, it is often not sufficient, and when the canals are filled with water it is probable that there will be a lot of floating debris at the front of the wave. Where nothing is done to remove the floating debris at key locations, the system runs the risk of creating some plugs and spills. The requirements for this type of operation depend on the duration of the nonflowing period—the longer is the period, so the greater is the need for resources and pre-season actions.

The closure of a canal must always be progressive. A too rapid drop in water level in an earthen canal is likely to generate scourges in the banks. The literature indicates some maximum recommended decreases in canal velocity.

The refilling of canals after a short break caused by a short-term event such as rainfall must be handled carefully as the demand for water may be uncertain. There is a need to carefully monitor water delivery vs any changes in the demand in order not to have to spill excessive volumes of water. There is also the risk that widespread and heavy precipitation will contribute to more uniform (or near-uniform) soil moisture levels in the service area and generate a new pattern of the demand with all the requests coming at the same time instead of in rotation as before.

References

Merriam, J.L. and Keller, J., *Farm irrigation system evaluation: A guide for management*, Department of Agricultural and Irrigation Engineering, Utah State University, Logan, Utah., 1978.

5

Design of On-demand Irrigation Systems

Large distribution irrigation systems have played an important role in the distribution of scarce water resources that otherwise would be accessible to few. Also they allow for a sound water resource management by avoiding the uncontrolled withdrawals from the source (groundwater, rivers, etc). Traditional distribution systems have the common shortcoming that water must be distributed by some rotation criteria that guarantees equal rights to all beneficiaries.

The inevitable consequence is that crops cannot receive the water when needed and reduced yields are unavoidable. However, this compromise was necessary to spread the benefits of a scarce resource. Among the distribution systems, the pressurised systems have been developed during the last decades with considerable advantages with respect to open canals. In fact, they guarantee better services to the users and higher distribution efficiency. Therefore, a greater surface may be irrigated with a fixed quantity of water.

They overcome the topographic constraints and make it easier to establish water fees based on volume of water consumed because it is easy to measure the water volume delivered. Consequently, a large quantity of water may be saved since farmers tend to maximise the net income by making an economical balance between costs and profits. Thus, because the volume of water represents an important cost, farmers tend to be efficient with their irrigation.

Operation, maintenance and management activities are more technical but easier to control to maintain a good service.

Since farmers are the ones who take risks in their business, they should have water with as much flexibility as possible, i.e., they should have water on-demand. By definition, in irrigation systems operating on-demand, farmers decide when and how much water to take from the distribution network without informing the system manager. Usually, on-demand delivery scheduling is more common in pressurised irrigation systems, in which the control devices are more reliable than in open canal systems.

The on-demand delivery schedule offers a greater potential profit than other types of irrigation schedules and gives a great flexibility to farmers that can manage water in the best way and according to their needs. Of course, a number of preliminary conditions have to be guarantee for on-demand irrigation. The first one is an adequate water tariff based on the volume effectively withdrawn by farmers, preferably with increasing rates for increasing water volumes. The delivery devices (hydrants) have to be equipped with flow meter, flow limiter, pressure control and gate valve. The design has to be adequate for conveying the demand discharge during the peak period by guaranteeing the minimum pressure at the hydrants for conducting the on-farm irrigation in an appropriate way.

In fact, one of the most important uncertainties the designer has to face for designing an ondemand irrigation system is the calculation of the discharges flowing into the network. Because farmers control their irrigation, it is not possible to know, a-priori, the number and the position of the hydrants in simultaneous operation. Therefore, a hydrant may be satisfactory, in terms of minimum required pressure and/or discharge, when it operates within a configuration but not when it operates in another one, depending on its position and on the position of the other hydrants of the configuration.

For on-demand irrigation, the discharge attributed to each hydrant is much greater than the duty. It means that the duration of irrigation is much shorter than 24 hours. As a result, the probability to have all the hydrants of the network simultaneously operating is very low. Thus, it would not be reasonable to dimension the network for conveying a discharge equal to the sum of the hydrant capacities. These considerations have justified the use of probabilistic approaches for computing the discharges in on-demand irrigation systems. Important spatial and temporal variability of hydrants operating at the same time occur in such systems in relation to farmers' decision over time depending on the cropping pattern, crops grown, meteorological conditions, on-farm irrigation efficiency and farmers' behaviour.

This variability may produce failures related to the design options when conventional optimization techniques are used. Moreover, during the life of the irrigation systems, changes in market trends may lead farmers to large changes in cropping patterns relatively to those envisaged during the design, resulting in water demand changes.

In view of the changes in socioeconomic conditions of farmers, a change in their working habits over time should not be neglected. Therefore, both designers and managers should have adequate knowledge on the hydraulic behaviour of the system when the conditions of functioning change respect to what has been assumed. Improving the design and the performance of irrigation systems operating on-demand requires the consideration of the flow regimes during the design process. It requires new criteria to design those systems which are usually designed for only one single peak flow regime.

Complementary models for the analysis and the performance criteria need to be formulated to support both the design of new irrigation systems and the analysis of existing ones. In fact the first performance criterion should be to operate satisfactorily within a wide range of possible demand scenarios. For existing irrigation systems, the models for the analysis may help managers in understanding why and where failures occur. In this way, rehabilitation and/or modernisation of the system are achieved in an appropriate way.

Design A Distribution Irrigation System

An irrigation system should meet the objectives of productivity which will be attained through the optimization of investment and running costs. A number of parameters have to be set to design the system. These parameters may be classified into environmental parameters and decision parameters. The environmental parameters cannot be modified and have to be taken into account as data for the design area. The latter depend on the designer decisions.

The most important environmental parameters are:

— climate conditions

— pedologic conditions

— agricultural structure and land tenure

— socioeconomic conditions of farmers

— type and position of the water resource

Information on the climate conditions is required for the computation of the reference evapotranspiration. Rainfall is important for the evaluation of the water volume that may be utilised by the crops without irrigation. Information on the pedologic conditions of the area under study is important to identify the boundary of the irrigation scheme, the percentage of uncultivated land, the hydrodynamic characteristics of the soil and the related irrigation parameters.

The water resources usually represent the limiting factor for an irrigation system. In fact, the available water volume, especially during the peak period, is often lower than the water demand and storage reservoirs are needed in order to satisfy, fully or partially, the demand. Also the location of the water resource respect to the irrigation scheme has to be taken into account because it may lead to expensive conveyance pipes with high head losses. Finally, the socioeconomic conditions of farmers have to be taken into account. They are important both for selecting the most appropriate delivery schedule and the most appropriate onfarm irrigation method.

All the above parameters have great influence on the choice of the possible cropping pattern. The most important decision parameters are:

— cropping pattern
— satisfaction of crop water requirements (partially or fully)
— on-farm irrigation method
— density of hydrants
— discharge of hydrants
— delivery schedule

The cropping pattern is based on climate data, soil water characteristics, water quality, market conditions and technical level of farmers. The theoretical crop water requirements is derived from the cropping pattern and the climatic conditions. It is important to establish, through statistical analysis, the frequency that the crop water requirement will be met according to the design climatic conditions. Usually, the requirement should be satisfied in four out of five years. The requirements have to be corrected by the global efficiency of the irrigation system.

The computed water volume has to be compared with the available water volume to decide the irrigation area and/or the total or partial satisfaction of the crops in order to obtain the best possible yield. The water requirements should account for the peak discharge. The designer needs updated maps at an appropriate scale (1:25 000, 1:5 000, 1:2 000) with contour lines, cadastral arrangement of plots and holdings. In fact, it may happen that a holder has two or more plots and might be served by only one hydrant located in the most appropriate point. The maps should allow for drawing of the system scheme.

The number of hydrants in an irrigation system is a compromise. A large number improves operation conditions of farmers but it makes for higher installation costs. Usually, for an appropriate density of hydrants it is better to plan no less than one hydrant of 5 l s-1 for 2.5 ha and, in irrigation schemes where very small holdings are predominant, no more than three or four farmers per hydrant. These limits will allow a good working conditions of farmers. Also the access to the hydrants should be facilitated. For this reason, in the case of small holdings it is appropriate to locate hydrants along the boundary of the plots. In case of large holdings it may be more appropriate to put hydrants in the middle of the plot in order to reduce the distance between the hydrant and the border of the plot.

The successive steps for designing an irrigation system include defining the network layout and the location of the additional works, like pumping station, upstream reservoir, and equipment for protection and/or regulation, if required. It is important to stress that the above phases are drawn on the maps. Because they are often not updated, field verification is needed in order to avoid passing over new structures that have not been reported on maps. If everything is done well (usually it never occurs), otherwise adjustments have to be done for one or more of the previous steps.

Computations of the discharges to be conveyed, the pipe diameters of the network, the additional works, like pumping station, upstream reservoir, and equipment for protection and/or regulation, are performed. The development process of an irrigation system follows a systematic chronological sequence represented in Figure 1.

Figure 1 The chronological development process of an irrigation system

When this process is a "one-way" process, obviously management comes last. However, experience with many existing irrigation schemes has proven that management problems are related to design. This is because the designer does not necessarily have the same concerns as the manager and the user of a system. It appears beneficial to consider the process in Figure 1 as a "whole", where the three phases are intimately interrelated (Figure 2). For these reasons, before moving on the construction of the system, models have to be used to simulate different scenarios and possible operation conditions of the system during its life.

Figure 2. Sound development process of an irrigation system

The simulation models will allow analysis of the system and will identify failures that may occur. Then the construction may start. After the construction, the designer should monitor the system and collect data on operation, maintenance and management phases. It will allow performing the analysis under actual conditions and will allow calibrating, validating and updating existing models, besides formulating new models, too.

Structure and Layout of Pressure Distribution Networks

Pressure systems consist mainly of buried pipes where water moves under pressure and are therefore relatively free from topographic constraints. The aim of the pipe network is to connect all the hydrants to the source by the most economic network. The source can be a pumping station on a river, a reservoir, a canal or a well delivering water through an elevated reservoir or a pressure vessel. In this publication, only branching networks will be considered since it can be shown that their cost is less than that of looped networks. Loops are only introduced where it becomes necessary to reinforce existing networks or to guarantee the security of supply.

On-demand Irrigation Network

Layout of Hydrants

Before commencing the design of the network the location of the hydrants on the irrigated plots has to be defined. The location of the hydrants is a compromise between the wishes of the farmers, each of whom would like a hydrant located in the best possible place with respect to his or her plot, and the desire of the water management authority to keep the number of hydrants to a strict minimum so as to keep down the cost of the collective distribution network.

In order to avoid excessive head losses in the on-farm equipment, the operating range of an individual hydrant does not normally exceed 200 metres in the case of small farms of a few hectares and 500 metres on farms of about ten hectares. The location of hydrants is influenced by the location of the plots. In the case of scattered smallholdings, the hydrants are widely spaced (e.g. at plot boundaries) so as to service up to four (sometime six) users from the same hydrant. When the holdings are large the hydrant is located preferably at the center of the area.

Layout of Branching Networks

On-demand distribution imposes no specific constraints on the network layout. Where the landownership structure is heterogeneous, the plan of the hydrants represents an irregular pattern of points, each of which is to be connected to the source of water. For ease of access and to avoid purchase of rights of way, lay the pipes along plot boundaries, roads or tracks. However, since a pipe network is laid in trenches at a depth of about one metre, it is often found advantageous to cut diagonally across properties and thus reduce the length of the pipes and their cost. It involves the following three steps in an iterative process:

— "proximity layout" or shortest connection of the hydrants to the source;

— "120° layout" where the proximity layout is shortened by introducing junctions (nodes) other than the hydrants;

— "least cost layout" where the cost is again reduced, this time by shortening the larger diameter pipes which convey the higher flows and lengthening the smaller ones. The last step implies a knowledge of the pipe diameters.

Application of Pipe Network Optimization

A search for the optimal network layout can lead to substantial returns. An in-depth study of a network serving 1 000 ha showed that a cost reduction of nine percent could be achieved with respect to the initial layout. This cost reduction was obtained essentially in the range of pipes having diameters of 400 mm or more. In general it may be said that the field of application of network layout optimization mainly concerns the principal elements of the network (pipe diameters of 400 mm and upwards).

Elsewhere land tenure and ease of maintenance (accessibility of junctions, etc.) generally outweigh considerations of reduction of pipe costs. In support of this assertion it is of interest to note that in the case of a 32 000 ha sector, which forms a part of the Bas-Rhône Languedoc (France) irrigation scheme, pipes of 400 mm diameter and above account for less than twenty percent of the total network length. In terms of investment, however, these larger pipes represent nearly sixty percent of the total cost.

In the case of schemes where the land tenure has been totally redistributed to form a regular checkwork pattern of plots, the pipe network can follow the same general layout with the average plot representing the basic module or unit. The layout of the pipe network is designed so as to be integrated with the other utilities, such as the roads and the drainage system.

Optimization of Branching Networks

The method commonly used involves three distinct stages:

Stage 1: Proximity layout

The aim is to connect all hydrants to the source by the shortest path without introducing intermediate junctions here denominated nodes. This may be done by using a suitable adaptation of Kruskal's classic algorithm from the theory of graphs.

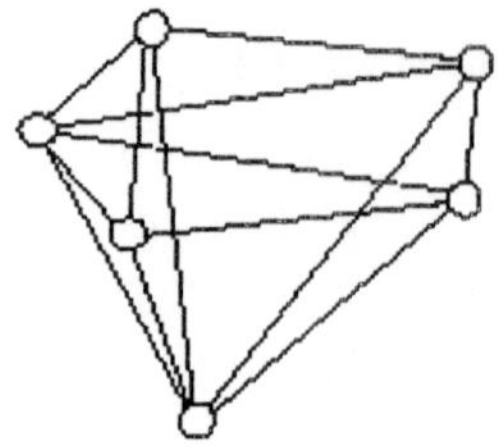

Figure 3. Proximity layout: application of Kruskal's algorithm

If a straight line drawn between hydrants is called a section and any closed circuit a loop, then the algorithm proposed here is the following: Proceeding in successive steps a section is drawn at each step by selecting a new section of minimum length which does not form a loop with the sections already drawn. The procedure is illustrated in Figure 3 for a small network consisting of six hydrants only. In the case of an extensive network, the application of this algorithm becomes impractical since the number of sections which have to be determined and compared increases as the square of the number of hydrants: $(n^2 - n)/2$ for n hydrants. For this reason it is usual to use the following adaptation of Sollin's algorithm. Selecting any hydrant as starting point, a section is drawn to the nearest hydrant thus creating a 2-hydrant subnetwork.

This sub-network is transformed into a 3-hydrant sub-network by again drawing a section to the nearest hydrant. This in fact is an application of a simple law of proximity, by which a sub-network of n-1 hydrants becomes a network of n hydrants by addition to the initial network. This procedure, which considerably reduces the number of sections which have to be compared at each step, is illustrated in Figure 4.

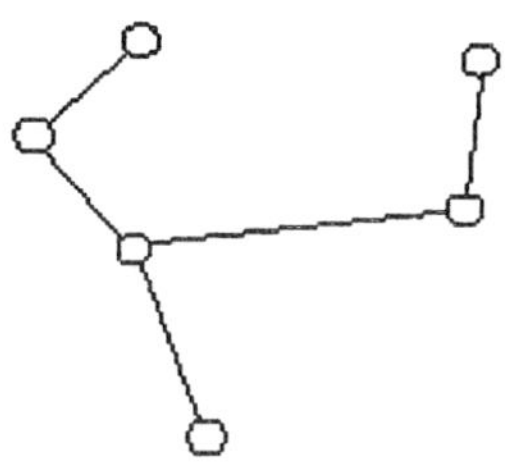

Figure 4 Proximity layout: application of Sollin's algorithm

Stage 2: 120° Layout

By introducing nodes other than the hydrants themselves, the proximity network defined above can be shortened:

Case of three hydrants

Consider a sub-network of three hydrants A, B, C linked in that order by the proximity layout. A node M is introduced whose position is such that the sum of the lengths (MA+MB+MC) is minimal. Let $\vec{i}\,\vec{j}\,\vec{k}$ be the unit vectors of MA, MB and MC and let dM be the incremental displacement of node M. When the position of the node is optimal then $d(MA + MB + MC) = (\vec{i} + \vec{j} + \vec{k})dM = 0$ This relation will be satisfied for all displacements dM when $\vec{i} + \vec{j} + \vec{k} = 0$ It follows therefore that the angle between vectors $\vec{i}, \vec{j}, \vec{k}$ is equal to 120°.

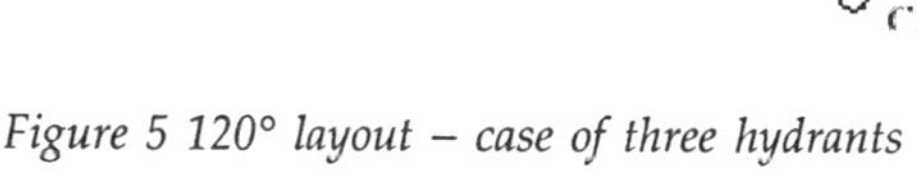

Figure 5 120° layout – case of three hydrants

The optimal position of the node M can readily be determined by construction with the help of a piece of tracing paper on which are drawn three converging lines subtending angles of 120°. By displacing the tracing paper over the drawing on which the hydrants A, B, C have been disposed, the position of the three convergent lines is adjusted without difficulty and the position of the node determined. It should be noted that a new node can only exist if the angle ABC is less than 120°.

When the angle is greater than 120°, the initial layout ABC cannot be improved by introducing a node and it represents the shortest path. Conversely, it can be seen that the smaller is the angle ABC, the greater will be the benefit obtained by optimising.

Case of four hydrants

The 120° rule is applied to the case of a four-hydrant network ABCD. The layout ABC can be shortened by the introduction of a node M_1 such that sections M_1A, M_1B and M_1C are at 120° to each other. Similarly the layout M_1CD is shortened by the introduction of a node M_1' such that M_1' M_1, M_1' C and M_1' D subtend angles of 120°. The angle AM_1M_1' is smaller than 120° and the node M1 is moved to M_2 by the 120° rule, involving a consequent adjustment of M_1' to M_2'. The procedure is repeated with the result that M and M' converge until all adjacent sections subtend angles of 120°. In practice, the positions of M and M' can readily be determined manually with the assistance of two pieces of tracing paper on which lines converging at 120° have been drawn.

Case of n hydrants

The above reasoning can be extended to an initial layout consisting of n hydrants. It can be shown that the resulting optimal layout has the following properties:

- the number of nodes is equal to or less than n-2;
- there are not more than three concurrent sections at any node;
- the angles between sections are equal to 120° at nodes having three sections and greater than 120° when there are only two sections.

In practice it is impractical to deal manually with the construction of a network consisting of four or five hydrants, involving the introduction of two or three adjacent nodes, even with the help of tracing paper. Several geometric construction procedures have been devised to facilitate such layouts, but these are rather cumbersome and the problem can only be resolved satisfactorily with the assistance of a computer. It is rarely necessary to create more than two or three consecutive nodes. Also, the benefit gained by optimising decreases as the number of adjacent sections increases.

Stage 3: Least-cost layout

Although the layout which results from applying the 120° rule represents the shortest path connecting the hydrants, it is not the solution of least cost since no account is taken of pipe sizes. The total cost of the network can further be reduced by shortening the larger diameter pipes which convey higher flows whilst increasing the length of the smaller diameter pipes which convey smaller flows. This will result in a modification of the angles between sections at the nodes. The least-cost layout resembles the 120° layout but the angles joining the pipes are adjusted to take into account the cost of the pipes.

The step which leads from the 120° layout to the least-cost layout can only be taken once the pipe sizes have been optimised. But this condition induces to a loop. In fact, for calculating the pipe sizes of the network, the layout should be already known. A method for the simultaneous computation of optimal pipe size and layout has been developed for particular distribution systems with parallel branches. Two different approaches have been adopted: the linear programming formulation and a special purpose algorithm. Both these two approaches have been applied to a simple example and their reliability and usefulness was demonstrated. Unfortunately, at this time, no commercial software packages are available for applying such method to actual networks.

Layout optimization methods

There is no doubt that the 120° layout is an improvement on the initial proximity layout and that the least-cost layout is a further refinement of the 120° layout. It is not certain however that the complete process produces the best result in all cases.

Usually, "rules of thumb" are applied by designers in selecting the best suitable layout and, later, optimization algorithms are applied for computing the pipe sizes. The optimum attained is relative to a given initial layout of which the proximity layout is only the shortest path variant. It could be that a more economic solution is possible by starting with a different initial layout, differing from that which results from proximity considerations, but which takes into account hydraulic constraints.

Several initial layouts of the network can be tested. The first of these should be the proximity layout. The others can be defined empirically by the designer, on the basis of the information available (elevation of the hydrants and distance from the source) which enables potentially problematic hydrants to be identified. By a series of iterations it is possible to define a "good" solution, if not the theoretical optimum.

References

Doorenbos, J. and Pruitt, W.O., *Guidelines for predicting crop water requirements*, FAO Irrigation and Drainage Paper 24. pp. 40-41, 1977.

D'Urso, G., Querner, E.P. and Morabito, J.A. "Integration of hydrological simulation models with remotely sensed data: an application to irrigation management", In: *Advances in Planning, Design and Management of Irrigation Systems as Related to Sustainable Land Use*, Vol. 2. J. Feyen *et al.* (eds.). Leuven, Belgium. pp. 463-472, 1992.

Menenti, M., Visser, T.N.M., Morabito, J.A. and Drovandi, A., "Appraisal of irrigation performance with satellite data and geo-referenced information", In: *Irrigation Theory and Practice*. J.R. Rydzewsky and K. Ward (eds.). London. pp. 785-801, 1989.

6

Mapping Water Networks

The canal and drainage systems within the service area need to be properly understood in order to develop appropriate and workable water management and operational strategies, ones that fully consider constraints as well as opportunities. In order to achieve this objective, it is necessary to map the flow routes and specify the flows in terms of timing, flow rates and volumes. Water accounting, also called water balance, refers to the accounting of all the influxes and outfluxes of water in a given space and time. It must consider all water that enters and leaves a defined area in a particular span of time. Thus any recycled water within the spatial boundaries is not included in the water balance.

It should take into account quantity and also water quality aspects, the use of lower quality water, and the impacts of agriculture practices on water resources. Although the concept of conjunctive use of irrigation water has been around for many decades, it is only in the last decade or so that systematic and comprehensive water balances have been carried out in major canal systems in order to gain a better understanding of the flows and water resources.

A thorough knowledge of all existing and potential sources of inflows and outflows in service area is required for efficient canal operation and good water management practices. This is done, in MASSCOTE, by assessing the hierarchical structure and the main features of the irrigation and drainage networks, natural surface streams and groundwater, and the mapping of the opportunities and constraints, including drainage and recycling facilities. Thus mapping of water network includes not only irrigation canal and drainage network but also any stream and/or natural channel and drain that crosses the service area and which interacts (or may interact in future) with the canal and drainage network and storage facilities. This also includes escapes that evacuate water to the drainage network.

The geographical distribution of the canal system and infrastructure within the service area is usually quite well known. However, this is often not the case for the drainage system, which typically has not been developed fully or has been developed in various phases, often with no reliable record of precise locations and of what has been maintained and is functional. Information on where and how much water is drained may help managers and decision makers in assessing the potential of recycling this water.

Also information regarding the layout and the (dry) discharges of streams that could be used for irrigation, if the popography allows, in water short system is often not available, thus managers do not consider water from these streams in allocation and distribution plan. Mapping of water network is particularly important for developing practical water management strategies and effective water resource allocation, in particular in (seasonal) water short system.

Foundation of Water Management

Water accounting should be considered as the foundation of water management and operation in the sense that it defines the requirements (demand) for surface water service. Through a spatially desegregated water balance it is possible to identify:

- various flows within the service area;
- the needs that the managers must satisfy; and
- opportunities for sharing the cost of operation among more users than only the farmers.

It also gives a good indication of the efficiencies of water management and allows the identification of environmental problems, such as waterlogging. Where done properly, a water balance can be used by managers to assess the conditions under which canal operations take place.

A comprehensive water balance is critical:

— at the initial project stage, for setting water services, designing appropriate management strategies and operational procedures;

— later, for appraising modernisation strategies to achieving updated performance targets.

Three important features for water balances are:

— Delineation of the physical boundaries: upper limit, lower limit and horizontal limit.

— Time frame: year, season, month, fortnight or ten-day period.

— Focus: water quantity and water quality.

Water quantities must account for all inflows and outflows, plus changes in internal storage. For water quality, the process is more complex and depends considerably on the biochemical and physical properties with time of the parameter under consideration. Along their paths through the water cycle, chemicals are absorbed, degraded, transformed, lost through aerial reaction, etc. Therefore, mass conservation applies to water but does not always apply easily to chemical constituents. Depending on the purpose, water accounting could be done on a seasonal or yearly basis at the entire scheme level and for submanagement units in order to facilitate management decisions.

Physical Units within the Service Area

There are many criteria to consider in defining the physical boundaries for an irrigation water balance. A water balance can be conducted for a field, a farm, a submanagement unit, an entire irrigation service area, and a river basin. Whatever the unit of evaluation, it is necessary to define upper, lower and horizontal boundaries of space. Spatial boundaries for a water balance to assist decisions regarding system management and operation would include:

— the gross service area of the project: often used as the first approach to examine a global water balance;

— canal hierarchy: main, secondary, tertiary and quaternary;

— institutional management: federation of WUAs, WUA, farmers organisation.

The above criteria can be included in the definition of the water balance. However, one of the more important aspects is pragmatism. Units for the water balance should be based on realistic boundaries for which flows can be either measured or estimated with reasonable accuracy. In an ideal situation, a water balance is conducted for the entire irrigation service area and each management subunit in order to allow the managers and operators to make decisions within their own subunits as well as at the entire project/ system level. However, whatever unit is chosen for the analysis, the boundaries must be clearly set and understood. Setting the spatial as well as the temporal boundaries for a water balance is very important. The failure to set these limits properly is often a main reason for errors made in computing water balances.

Temporal Boundaries

Temporal boundaries are critical when computing a water balance. Depending on the objectives for which the water balance is conducted, temporal limits can be set as multiple years, one year, six months, an irrigation season, monthly or fortnightly. For example, making long-term recommendations on the basis of only a one-year water balance is not recommended because such data are often not representative of normal conditions. The values of most of the water balance inputs, such as rain, surface allocations, and evapotranspiration vary from year to year. For the purpose of making long-term recommendations, 4-5-year average values from water balances done on a yearly basis must be considered. For the purpose of evaluating modernisation strategies, a time frame of a year, six months or a single irrigation season is advisable.

Monthly or fortnightly water balances are required where the objective is to use the values for real-time management decisions. However, it is often difficult to assess changes in the groundwater storage on a scale smaller than one year. Nevertheless, it is necessary for managers and operators of the irrigation system to keep an account of where water is coming from and where is it going to within the management units in order to be able to make efficient decisions regarding water conservation, allocation and distribution.

WATER BALANCE TERMS

Whatever the spatial unit under consideration, a number of basic flow parameters need to be evaluated (Figure 1):

- — irrigation diversions;
- — surface runoff into and out of the spatial boundary;
- — evapotranspiration (ET) from fields and other areas such as canals, drains and other non-irrigated areas;
- — rainfall within the spatial boundaries;
- — surface drainage, lateral groundwater flows and vertical drainage within the lower boundary limit.

Irrigation Diversions

Irrigation diversions are often measured through measurement devices at bifurcations. For a water balance, irrigation diversions entering the spatial boundary must be known.

Rainfall Gauges

Rainfall should be measured with sufficient density of points to account for the spatial variability of precipitation, especially if the time frame of events is short. It is common to see large irrigation projects covering dozens of square kilometres that

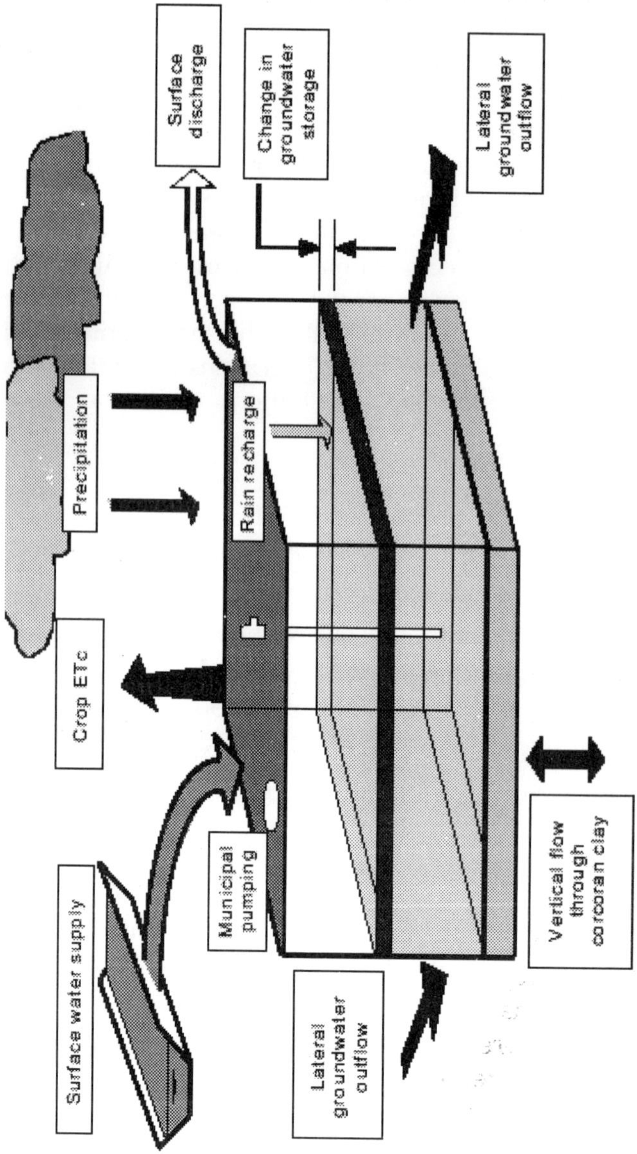

Figure 1. Water flows relevant for water accounting in agriculture

have only one rainfall gauge despite a high variability of rainfall during storm episodes. In practice, the number of rainfall gauges should be adjusted to the local spatial variability of the precipitation. A reasonable distribution is one gauge every 5 km in a medium-size system, and one every 10 km in a large system.

Crop Evapotranspiration

Evapotranspiration or crop water use is usually the largest and the most important component of water balance. It is obtained as the product of crop area and the estimation of crop evapotranspiration (ETc). However, where non-crop vegetation (trees, bushes, etc.) covers a non negligible part of the CA, its evapotranspiration also constitute a significant part of the water balance. Crop evapotranspiration or crop water requirement can be assessed by multiplying the reference evapotranspiration and the crop coefficient: $ETc = Kc \times ETo$; where: *ETc* is crop evapotranspiration; *ETo* is the reference evapotranspiration; and *Kc* is the crop coefficient.

The reference crop evapotranspiration represents the evapotranspiration from a standardised vegetated surface. The only factors affecting ETo are climate parameters. Thus, ETo can be computed from weather data. There are a number of methods for calculating ETo, but FAO recommends using the FAO Penman-Monteith method. Alternatively, the FAO CROPWAT programme can be used to assess both ETo and Etc.

Some weather data (temperature, solar radiation, relative humidity, and wind speed) are required in order to calculate ETo irrespective of the method used for its calculation. However, it is not always easy to obtain these data from the nearest meteorological station. Where no data are readily available, climate databases such as CLIMWAT (FAO) or the Climate Atlas of the International Water Management Institute (IWMI) provides values for weather parameters needed for *ETo* calculations. These source also provide good estimates of *ETo*. The crop coefficient, Kc, is basically the ratio of *ETc* to *ETo*, and it depends mainly on the crop variety and the growth stages of the crop. FAO provides *Kc* values for different crops and their growth stages, which are used widely throughout the world for estimating *ETc*.

The ETc calculated through above equation is the evapotranspiration from crops grown under optimal management and environmental conditions, with good water availability and no limitations of any other input.

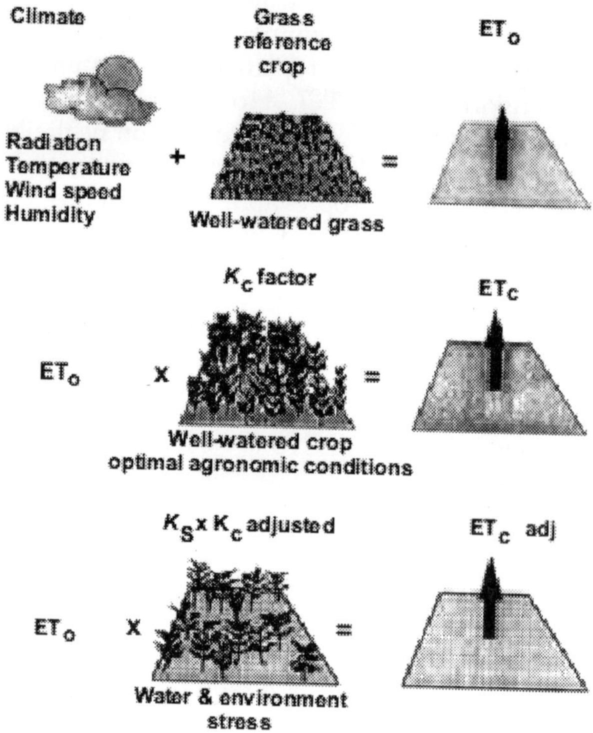

Figure 2. Reference ET, crop evaporation under standard (ETc) and non-standard (ETc adj) conditions

In most cases, actual crop water use differs from this potential *ETc* because of non-optimal conditions, such as the presence of soil salinity, water shortage, waterlogging, pests, etc. These conditions may reduce the evapotranspiration rate below *ETc*. In order to address this problem, a water stress coefficient, *Ks*, is introduced into the equation (Figure 2). *Ks* is dependent on the available soil moisture and ranges between 0 and 1. When the rootzone depletion is lower than water that is readily available to the crop, *Ks* = 1, meaning that water uptake to plants is equal to ETc where the soil is wet: ETc adj = $Ks \times Kc \times ETo$.

Drainage

Drainage should be measured at key points, particularly when it leaves the spatial boundary set for water balance, and possibly monitored for water quality if necessary. Water quality monitoring of drainage water from irrigated areas is important, in particular where this water is to be used for irrigation downstream, in order to keep a check on the safe levels of agrochemical loads.

Measurements of Groundwater

Groundwater fluxes, i.e. lateral flows and vertical drainage, are often the most difficult aspects to handle in a water balance. While direct measurements of groundwater flows are not possible, water-table levels can be measured, and groundwater or hydrological modelling can be used to reconstruct the flows using trial and error and comparison with field data. However, it is not always easy to calibrate these models owing to a lack of empirical data. An easier way of monitoring the changes in groundwater is to install monitoring wells, which could be made locally. Groundwater is an important component of water accounting.

The setting of the lower spatial boundary usually determines whether the use of (shallow and/or deep) groundwater is considered a supply or a mere recirculation of surface water supply and rainfall. However, a distinction between shallow groundwater and deep groundwater should be made in the water accounting procedures if they are to be conducted for assisting in management decisions, particularly in semi-arid and arid regions. In case shallow groundwater (less than 20 m) is incorporated in the lower spatial limits of the water accounting system, there is no need to take account of additional water supply from the tubewells fed by the shallow groundwater.

Only lateral subsurface flows entering and leaving the limits of the system should be considered. Generally, water extracted from deep groundwater (more than 20 m), is considered beyond the limits of the system. Therefore, the specific water supply from deep ground water is usually added to the inflow. In order to avoid double-counting of water entering into the spatial boundary of the water balance area, it is necessary to clearly identify the groundwater that is pumped within the spatial boundary and which may be considered as the recirculation of surface water and rainwater entering the spatial boundary. This groundwater should not be accounted for whereas groundwater that is pumped outside the spatial boundary but is used for irrigation within the boundaries of water balance should be taken into account as inflow/supply.

Confidence Intervals

A certain amount of error or uncertainty is inherent in all measurement or estimation processes. Therefore, the true or correct values for the water volumes needed to calculate terms such as "irrigation efficiency" are unknown. Estimates must be made of the component volumes, based on measurements or calculations. One method of expressing the uncertainty is to specify the confidence interval (CI) that is associated to the estimate of one value. If it is believed that a reasonable evaluation of data indicates that the correct value lies within 5 units of 70, then it should be stated that the quantity.equals 70±5.

More specifically, the meaning of a CI should be illustrated, e.g.: "The investigators are 95-percent confident that their estimate of the irrigated area in the project is within ±7 percent of 500 000 ha (between 465 000 ha and 535 000 ha)." Statistically, a CI is related to the coefficient of variation (CV), where: CV = (mean) / (standard deviation); CV has no units. In addition, CI = ±2 × CV, where the CI is expressed as a fraction (%/100) of the estimated value. Stated differently, if the CI is declared to be 0.10, this means that the ±2 standard deviations cover a range of ±10 percent of the stated value.

Assuming a normal distribution of data, then in about 68 percent of cases the true value is found within plus or minus one standard deviation of the estimated value. Similarly, in about 95 percent of cases (from which comes the "95-percent confident" statement), the true value is found within plus or minus two standard deviations of the estimated value.

Many terms in a water balance are the result of the addition or multiplication of individual terms and parameters. There are methods for calculating the CI of such aggregated parameters, such as $m = m_1 + m_2$ or $m = m_1 \times m_2$ when these parameters and terms are independent.

Closure of Water Balance

The closure of a water balance is a term or a set of terms of the balance that cannot be measured but have to be estimated from the assessment of the other terms. The closure of water balance is always computed with a high uncertainty because it accumulates all the uncertainties of the other terms. This aspect can be illustrated with three cases: perennial vegetation, groundwater, and seepages.

Non-crop vegetation in the CA might be an important term of the water balance as many trees thrive on water made available from surface irrigation. This phenomenon can be assessed visually by looking at areas within and outside the CA. In dry zones of the tropics, there is often a clear difference in terms of type of vegetation and foliar development, and, as a consequence, in terms of water consumption. However, it is not easy to estimate the areas covered and the unit consumption. The value of the perennial vegetation water consumption can be computed as the closure of the balance after having estimated all the other terms.

The lateral net contribution of groundwater (not including recharge from the canals and surface of the CAs) can be estimated as the closure of the other terms, and as such known with high uncertainty. In the case of the GLBC, with high inaccuracy on the other terms of the balance, the initial estimation is that the closure is probably known at ±55 percent. Thus, the groundwater lateral flows might be between 238 MCM and 732 MCM per year.

Seepage measurement is also a good example of how a closure of a water balance is fraught with uncertainty. Most experts recommend measuring seepage using ponding tests, from the records of drop in water level in an isolated reach (no inflow and no outflow) taking into account the evaporation. The other possible method, the inflow-outflow method is much less precise. It consists of measuring the flow in the canal at two locations and deducting all the inflows and outflows occurring between the two sections considered: Seepage = Inflow - Outflow. In order to provide an accurate estimate of seepage, the water lost should be significantly greater than the error in measuring the flows. The problem of accuracy can be illustrated through the case of the main canal in the GLBC project.

A measurement campaign was done to assess the seepage. Discharge was measured at 0 km and 50 km in order to estimate the losses through seepage. Inflow was 78.90 m^3/s, total outflow was 73.36 m^3/s, leading to a seepage estimate of 5.6 m^3/s for 50 km, i.e. 0.11 litres/s/ml. The problem is that the uncertainty about the inflows and outflows is such that this result is probably known plus or minus 200 percent. Even with an ideal situation where inflows and outflows are known in the field at 2.5-percent accuracy, the inflow is known thus ±1.97 m^3/s and the outflow ±1.83 m^3/s. In these circumstances, the seepage estimator is know at ±3.8 m^3/s, and thus lies between 1.8 m^3/s and 9.4 m^3/s.

Risk of Double Counting

The risk of double-counting in a water balance is real and it is always necessary to ensure that only the fluxes through the boundary of the system are accounted for. A typical example of this is groundwater pumping or recycling from drainage. While it is important to have an estimate of their value, they should not be counted in the water balance if this water has already been accounted for as inputs either from the rain or from the irrigation supply. Only external water to the system should be counted, which can be deep groundwater and lateral water from aquifers. The same applies for the surface recycling facilities.

Water Balances at Management Level

The rationale behind water balances may differ depending on whether the goal is to use the overall water balance within the gross service area or to evaluate the management and operations of potential modernisation strategies. Water balances can be useful to the management of a canal system in several ways:

- to set up an efficient and user-oriented water management strategy;
- to manage water in real time during the season and operate the system accordingly;
- to assess performance in water management as well as in delivering the service by providing an estimate of external indicators.

Water balances are vital to understanding issues such as the potential for improvement (e.g. water savings), the different uses of water within the service area, and, particularly, non-crop water use. In Cabannes, France, a modernisation project early in the 1980s was designed explicitly on the basis of a water balance. The scheme was divided into two sections:

- The upstream part, devoted mainly to cereals and field crops, was modernised using modern surface irrigation technologies (mainly furrow irrigation) to maintain high the recharge of groundwater for the downstream part of the system as well as for some domestic supply.
- The downstream part, devoted mainly to orchards, was modernised with drip irrigation using shallow groundwater pumping stations. The water balance of the whole system was checked carefully in order to ensure a sustainable supply for both sections, in the

knowledge that the modernised downstream part would need to divert much less water with localised irrigation technique compared with the earlier surface techniques.

Water Accounting for Institutional Arrangements

In many cases, canal systems provide water services to different types of users, regardless of the fact that the main objective is to supply water to crops. Multiple uses of water are the common rule and not an exception.

Water for Crop Use and Water for Other Uses: Although most irrigation systems have been built solely to supply water to crops during dry periods, in practice some of them have been feeding other uses of water, from the management losses or natural seepages. This is often the case for rice systems, where water ponding generates high shallow groundwater flows that might be tapped by others users. In old, gravity-fed systems in southeast France, figures of less than 25 percent of annual surface water supply for irrigated crops are not unusual, the remaining fraction being shared by groundwater recharge and surface-stream supply. This phenomenon is not well documented in the irrigation systems in developing countries. However, one such example is the Kirindi Oya irrigation system in Sri Lanka, where water for natural vegetation and homestead gardens in tropical humid areas takes an important fraction of the irrigation water input.

In these types of systems, water accounting may have a strong impact on the identification of different uses of water, and to a lesser extent on the qualification of users or beneficiaries that take advantage of water management. It is the basis on which managers, users and beneficiaries can discuss management strategies within the project.

Water accounting for canal operation is also useful for performance assessment within a period of time (ten days, month, season and year). In particular, it is useful in comparing water deliveries with water uses. The water balance of the Kirindi Oya system (for one full year (1998), considering the two crop seasons as well as the fallow period. This water balance was carried out because of the presumed poor performance of irrigation management. The fact was that the water duty (water delivered for irrigation from the main reservoirs divided by the CA) was dramatically high, values of 3 000-4 000 mm/season were not unusual, this was the initial motivation for investigating the matter. The water balance has changed completely the way of looking at performance in this project. The striking facts that were brought to light in 1998 were:

— Crop evapotranspiration accounts only for 23 percent of the total water supply (irrigation plus rainfall).

— The bulk of the consumption lies in homestead gardens and coconut trees, fed mostly by lateral flows from the irrigated areas, and which are very beneficial for the people.

— Other users are those who fish in the tanks, and cattle growers for their use of the fallow period on paddy-fields.

— A win-win situation was identified for the lagoon, where excessive freshwater from irrigated areas are generating negative impacts for a total of 3 percent of the total water volume.

— The real water losses (at the sea mouth, where no more value can be assigned to freshwater in the project) account for 16 percent.

— The potential for water savings (16 percent + 3 percent) is significant compared with crop use (23 percent) although it is necessary to consider that part of these water losses occur at flood times and would be difficult to value. A water balance can also provide estimates of external indicators, such as irrigation efficiencies, ratio of relative water supply (water required vs total water available), and crop yield per unit of water supplied.

Water Accounting for Canal Operation

Water accounting can also be important for real-time

decisionmaking for adjusting operation and upstream deliveries. On this short time scale, it is more a combination of indicator assessment and water accounting. For example, the presence of excessive drainage flows downstream from a subarea can indicate that there is too much water entering the CA compared with the current use. Observing this can be a trigger for action. The managers need to know:

— by how much to reduce the inflow to the CA in order to reduce significantly the drainage without creating a water shortage in the downstream part of the CA; and

— how long it takes for an inflow change to be reflected in the drainage flow. These parameters of the reaction to the presence of drainage flow can be adjusted by trial and error, and simple water accounting.

Development of Water Accounting Process

The uncertainty relating to water accounting within a gross command area can be reduced progressively with time. The compilation of data over long periods of time allows the uncertainty on some parameters to be reduced, and some inconsistencies to be detected and corrected. This improves water balance data and helps to narrow the gap between estimations and actual values. Improvements in water balance data require good measurement devices and efficient information management systems, which cost resources in terms of time and money. The intelligent use of the memory of water accounting enables improved decisiontaking in the following seasons and years.

Quality of Water

The quality of water in irrigation is also an important issue for the environment, resource management, and the health of the local population. A separate account of quality of water entering into and leaving out of a physical boundary helps in identifying water related environmental hazards. This information then could lead to the identification of appropriate mitigation strategies. The main issues of water quality in an irrigation system are related to:

— salinity - reduced crop yield, reduced soil quality;

— environmental pollution - disposal of industrial and municipal wastes into irrigation canals;

— drainage water from irrigated area with agrochemical loads;

— health - water related diseases, arsenic and heavy-metal contamination.

There are various impacts to consider in relation to the use of marginal quality irrigation water.

— Soil pollution/contamination: Marginal quality irrigation water can affect crop yields severely and damage soils. In particular, in semi-arid and arid countries (e.g. Egypt and Pakistan), soil salinity and sodicity are major problems that have been exacerbated by irrigation from saline groundwater because of unreliable surface water supplies. High levels of heavy metals in the water are likely to accumulate in the topsoil and then enter the food chain.

— Conflict with other uses: Wastewater from small industries as well as municipal waste is frequently discharged into the canals and surface water streams. This creates pollution and health hazards as these canals often provide water for drinking purposes and domestic use.

Canals running through settlements, villages and urban areas are also frequently used as dumping grounds for refuse. This creates pollution and health hazards for the adjacent communities. It also causes problems of water conveyance by blocking the canals, and eventually disrupts water distribution downstream.

While water-borne diseases are caused by consuming contaminated water, stagnant water in waterbodies, canals and fields are major sources of vector-borne diseases as they become breeding grounds for insect vectors, especially mosquitoes. The uptake by plants of heavy metals and arsenic through direct contact with irrigation water or through accumulation in the soil also poses a threat to human health as these elements can enter the food chain.

Water quality requires its own M&E system, which is not always possible for irrigation managers to organise and handle because of the lack of technical and financial resources. However, a minimum dataset of water quality indicators needs to be developed and monitored in canal systems, in particular for those providing water for multiple uses and where water quality is a major issue, e.g. where the water is known to be saline/ sodic, or contaminated with high levels of arsenic and/or heavy metals.

References

Busman, J.D., *Optimizing control of surface irrigation using concurrent evaluation of infiltration*, PhD Dissertation, Agricultural and Irrigation Engineering, Utah State University, Logan, Utah. Unpublished document. 209p.

Doorenbos J. and W.O. Pruitt, *Crop water requirements*, Irrigation and Drainage Paper 24, Rome. 144p, 1977.

Garton, J.E., *Designing an automatic cut-back furrow irrigation system*, Oklahoma Agricultural Experiment Station, Bulletin B-651, Oklahoma State University, Stillwater, Oklahoma, 1966.

Gharbi, A., *Effect of flow fluctuations on free-draining and sloping furrow and border irrigation systems*, MS Thesis, Agricultural and Irrigation Engineering. Utah State University, Logan, Utah., Unpublished document, 123p, 1984.

7

Drainage Pipes and Accessories

For many years, clay and concrete pipes were predominantly used until the introduction of smooth plastic drainpipes around 1960. Soon afterwards corrugated plastic pipes came into common use. Clay, concrete and plastic pipes give satisfactory results if they meet quality standards and are properly installed. Collector pipes are made of concrete or plastic.

Pipes that are manufactured from the latter type of material are not yet competitive for diameters exceeding 200 mm. However, perforated corrugated plastic collectors, wrapped with a sheet envelope, may be installed comparatively easily if the surrounding soil consists of quicksand or has other "quick" properties. Once installed, the collector can act as a drain, cancelling the quick condition of the soil and facilitating the connection of laterals and/or the installation of manholes. In theory, there are valid considerations to select specific types of drainpipe. In practice, selection is mostly based on cost comparison and on local availability. In addition, the following observations may be relevant:

— If all types of pipe are available, the use of corrugated plastic pipes has distinct advantages.

— If pipes are not locally available, local manufacture of concrete pipes is the most straightforward and the easiest to implement. It requires less skill than manufacturing other types of pipe, and is already economical on a small scale. Plastic pipes occupy an intermediate position: local manufacturing from imported raw material is indeed possible for reasonably large quantities.

— Plastic pipes are particularly suited for machine installation. They have the advantage that their performance is the least affected by poor installation practice.
— The manufacturing cost of small diameter pipe is usually of the same order for clay tiles, concrete tiles and plastic. For large diameter pipes, however, concrete is usually the cheapest and plastic the most expensive.

Clay Tiles and Concrete Tiles

Clay tile may be either porous or glazed. Pipe sections are abutted against each other and water enters through the joints. The porous type usually has butt joints, but it may also have flanges (also referred to as 'collars' or 'bell joints'). The latter type of tile is more expensive, and the extra cost is only justified in very soft soils. Good quality pipes are adequately baked and are free from cracks and blisters. Clay tile with cracks or other visible shortcomings and badly formed pipes should not be used. Standard drainpipe sizes are 50, 65, 75, 80, 100, 130, 160, and 200 mm inside diameter.

In the United Kingdom, the nominal minimum size is 75 mm inside diameter, which has a generous capacity to carry water, and thus diameter is rarely a significant design consideration when using clayware pipes for laterals. The wall thickness varies from 12 to 24 mm, and may be expressed as 0.08d + 8 mm, where d is the inside pipe diameter in mm. Current clay tiles have lengths of 300 or 333 mm, yet in some countries greater lengths are available. In Germany, clay tiles were provided with longitudinal grooves at the outside wall, facilitating water flow alongside the drain in combination with envelope materials.

Clay tile is very durable and highly resistant to weathering and deterioration in aggressive soil conditions e.g. in soils containing sulphates and corrosive chemicals. It can be used in almost all circumstances. Clay tile is lighter than concrete and has excellent bearing strength. It is however fragile (especially the German grooved type), and must be handled with care. Clay tiles require a good deal of manual handling, although manufacturers have improved this by various methods of bulk handling.

Manufacturing of clay tiles requires a great deal of skill and a well-equipped plant. The major quality features are straight joints, absence of cracks and homogeneity of the raw material (well-mixed clay). The maximum water absorption rate after being immersed in water for 24 hours should be less than 15 percent of the weight of the tile. The weight of 1 000 tiles should exceed certain minimum values, e.g. 1 400 kg for 60 mm diameter pipe and 2 000 kg for 80 mm diameter pipe.

In some areas, clay and concrete tiles are still laid manually in a hand dug or mechanically excavated trench. These pipes may be covered with bulky materials or with 'envelopes' in strip form. Clay tiles should be installed in such a way that a perfect alignment between individual pipes is obtained. The maximum gap between individual pipes may not exceed 3 mm, except for sand where it should be not more than 2d85, i.e. the particle size for which 85 percent of the soil particles on dry weight basis have a smaller diameter.

Concrete Tiles

Concrete tile has been used on a large scale, e.g. in Egypt, Iraq and other countries. It is used if clay tile is not available, or if greater diameters must be applied. Concrete pipes are used mostly in medium to large sizes, with inside diameters of 100, 150 and 200 mm and up, and section lengths of 0.60, 0.91, 1.22 and 2.40 m. Tile over 300 mm inside diameter is usually reinforced. Butt joints are common. The manufacture of concrete tiles is much simpler than that of clay tiles. Pipes should be well formed, finished, free from cracks and chips, and properly cured. Concrete pipes should be used only when soil and groundwater analyses have established that conditions are suitable for their use.

Pipes made with ordinary cement are liable to deteriorate in acidic and high sulphate soils, and by water carrying certain alkali salts or other chemicals. Concrete pipes should not be used at locations where industrial waste or house refuse has been collected. Special high sulphate-resistant cements and high density concrete should be used to resist chemical attack. Concrete pipes may disintegrate slowly from weathering, and are subject to erosion from fast flowing water carrying abrasive material. However, under a wide range of conditions, a permanent installation is lasting and justified.

Advantage of Plastic Pipes

The main advantage of plastic pipes over clay and concrete pipes is their low weight per unit length, greatly reducing transportation cost. An additional cost saving factor is the reduced need for the labour, required for installation. Smooth plastic pipes were made of rigid polyvinyl chloride (PVC) and were provided with longitudinal slits to permit water entry. Smooth plastic pipes have never found a widespread use because they were rapidly superseded by corrugated pipes, that became available in 1963. They were so successful that they gradually started to replace clay and concrete pipes. This process is still continuing in various countries. The corrugated shape of the wall makes the pipe not only flexible, but also more resistant to compression than the smooth pipe, for the same quantity of raw material.

The introduction of corrugated pipe was a milestone in the history of agricultural drainage. This flexible pipe is very well suited for mechanised installation. Hence, the installation costs are significantly reduced. In addition, corrugated pipe has facilitated the development of trenchless installation techniques. The switch from clay and concrete pipes to corrugated plastic pipes was expected because corrugated plastic pipes were advantageous, viz.:

— Light weight makes handling easier, even for great lengths.
— Long, continuous length eases handling, gives less alignment problems, and reduces stagnation of pipe supply resulting in a high installation rate for drainage machines.
— Flexibility and coilability facilitate handling, transportation and installation.
— Greater and more uniformly distributed perforation area, facilitating access of water.
— Easy wrapping with envelope materials.
— Safer implementation without too wide joints or misalignment.
— Less labour intensive and consequently lower labour cost for manufacture, handling, transportation and installation.
— Inert to all common soil chemicals.

Corrugated pipes also have disadvantages, compared to clay and concrete pipes:

— Vulnerability to deterioration from UV-radiation when exposed to sunlight for long periods, especially if made of PVC.
— Increased brittleness at low temperatures.
— Increased deflection risk at high temperatures and excessive stretch during installation.
— Lower deflection resistance under permanent load.
— Risk of collapse under sudden load, e.g. by trench wall caving or stones.
— Smaller transport capacity for the same inner diameter because of corrugation roughness.
— Not fire resistant.
— Not easy to relocate in the field with a tiie probe without damaging the pipe.

Corrugated plastic drains are made of PVC, high-density polyethylene (PE) and polypropylene (PP). Preference of one of these materials is based on economic factors. In Europe, corrugated drains are mainly made of PVC except for the United Kingdom, where they are made of PE and a minority of PP. In the United States and Canada, most drainpipes are made of PE, largely because of the low price of the raw material. Good quality pipes can be made of both PVC and PE although these raw materials have different physical properties:

- The lower stiffness of PE means that pipes may be easily deformed under load, especially at temperatures approaching 40°C, and if they are subjected to longitudinal stress.
- PVC pipes are more susceptible to UV-radiation and become brittle at exposure; storage of unprotected pipes in the open should therefore be avoided.
- PVC pipes are more brittle at low temperatures than PE pipes; PVC pipes should not be installed at outside temperatures below 3°C because the risk that cracks will be formed is too high.
- PVC softens at 80°C and drainpipes will deform when exposed to such temperature. Especially in arid and semiarid areas, special care shall be taken to prevent such storage conditions.
- PVC has some environmental disadvantages: it forms hydrochloric acid when burnt.

In northwest Europe, PP pipes have been introduced for agricultural purposes. They are not widely used, but they are quite suitable for application in greenhouses, because they are heat resistant and tolerate disinfection of soils by steam vapour. European pipe sizes usually refer to outside diameter. Standard outside diameters are 40, 50, 65, 80, 100, 125, 160 and 200 mm. Larger diameters are available as well. North American pipe sizes refer to inside diameters, which are 102, 127, 152, 203, 254, 305, 381, 457 and 610 mm. The inside diameter is normally 0.9 times the outside diameter. Corrugated plastic pipes of not too large a diameter (up to 250 mm) are delivered in coils. Larger diameter pipes are supplied in lengths of 6 m.

Water enters corrugated pipes through perforations, which are located in the valleys of the corrugations. Elongated openings or 'slots' are common, yet circular openings may be found as well. The perforations may have a diameter or slot width usually ranging between 0.6 to 2 mm. The length of the slots is approximately 5 mm, but sawn slits of larger diameter pipes may be longer. The perforations should be evenly distributed over the pipe wall, usually in at least four rows with a minimum of two perforations per 100 mm of each single row.

In Europe, the perforation area should be at least 1200 mm^2 per metre of pipe. Machine installation of corrugated plastic drainpipes is very straightforward. Smaller diameter pipes are usually carried on a reel on the machine and wound off while the installation proceeds. Larger diameter pipes are mostly laid out in the field and guided through the machine. A thorough control of the pipes and a careful installation is nevertheless always necessary to prevent pipe damage and longitudinal stretching. Regular quality control of corrugated plastic pipes is very important. The impact of sudden loads, simulating trench wall caving on the pipe at temperatures corresponding to the ambient installation temperature should be part of a testing programme.

Pipe Accessories

Subsurface drainage systems require accessories and special structures such as pipe fittings (couplers, reducers, junctions, end caps), gravity or pumped outlets, junction boxes, inspection chambers (manholes), drain bridges, non-perforated rigid pipes, blind inlets, surface inlets, controlled drainage or subirrigation facilities, and cleaning provisions. Some fittings are made by pipe manufacturers, others are manufactured by specialised companies, and others are fabricated on the spot.

Effect of End Caps

End caps prevent the entrance of soil at the upstream drain-end opening. They can be made of the corresponding pipe material but any other durable flat material can be used for this purpose as well.

Couplers

Corrugated pipes generally have external 'snap-on' couplers to connect pipes of the same diameter. Alternatively, a piece of pipe of the same diameter that is split for easy placing around both pipe ends, and firmly wrapped with tape or wire to keep it in place during installation, can be used instead. Internal couplers can be used with the trenchless technique to prevent separation of connected pipes when passing through the pipe feeder device. Pipes can also be connected internally by making a slit in the end of the upstream pipe and forming a cone that is pushed into the end of the downstream pipe. Such connections are not very reliable and do impede the discharge of water and suspended solids.

Reducers

Reducers connect two pipe ends of different diameters.

Range of Pipe Fittings

A wide range of pipe fittings, made of various raw materials, is commercially available for all kinds of pipes. Fittings for clay, concrete and corrugated plastic pipes are generally made by the various pipe manufacturers and therefore they are mostly not interchangeable. Cross, T and Y-pieces connect laterals or collectors with collectors. Many fittings are fabricated with multiple sizes at the ends facilitating the connection of various sizes of collectors and laterals. The end sides of the fittings are cut off, or adapted by removing some parts in the field to attach to the appropriate diameter. Simple connections with elbows and T-pieces on top of the collector are nowadays used to connect laterals with collectors.

Influence of Protection Structures

Drain Bridges

The undisturbed natural soil in which the pipes are laid normally has enough strength to support the pipe. However, when the drain crosses a soft spot where the soil has not yet settled, e.g. a filled-in former ditch, drain bridges should be used to maintain the level of the drain during settlement of the soil. Drain bridges can be made of timber blocks on which the drain is laid or of a continuous length of solid, rigid pipe surrounding the drain.

Rigid Pipes

Drainpipes can be connected to or slid into a rigid, reinforced concrete, plastic or coated steel pipe where they have to cross a road, a waterway, a gutter, unstable soil, a row of trees to prevent roots from growing into the pipes, or other obstacles.

Importance of Inlets

- *Blind Inlets:* Blind inlets are intended to drain stagnant pools, while sediments are intercepted. They consist of a trench above a drain that is filled with porous material. Durable material, such as stones, gravel and coarse sand is preferred as trench backfill. The gradation may vary from finer material at the surface to coarser with depth, although the trench can also be filled with one suitable porous material. The advantage of blind inlets is the initial low costs and the lack of interference with tillage operations. However, in general the use of blind inlets has been unsatisfactory because they tend to clog at the surface with fine soil particles and other sediments.
- *Surface Inlets:* Surface water inlets are incidentally used to evacuate surface water from localised areas through the drainage system when the construction of ditches is not feasible or impractical. A proper silt trap is essential to prevent or reduce drain siltation. The open inlet can be in the collector line although it is better

located next to the collector and connected to it with a siphon as a safeguard against poor maintenance. Surface inlets are usually made of masonry or cast-in-place concrete, but concrete and rigid plastic pipes can also be used. A metal grating is usually installed to restrict the entry of trash and waste.

Connection Structures

Junction Boxes: Junction boxes are used where two or more drains (laterals and/or collectors) come together or where the diameter or the slope of the collector changes. They can be pre-casted or made of masonry or cast-in-place concrete, but also rigid plastic or concrete pipes can be used for this purpose. Junction boxes can be combined with a silt trap and extended to the soil surface. The bottom of the silt trap should be at least 0.30 m below the bottom of the inlet of the downstream pipe. The invert of the entering laterals should be positioned at least 0.10 m above the top of the leaving collector to further sedimentation in the silt trap.

Blind junction boxes will not hinder field works. The lid should therefore be situated at a minimum depth of 0.40 m below soil surface. They can be exposed if inspection and occasional cleaning is required. With the lid at the soil surface, the junction box is not so very much different from an inspection chamber, yet it hampers field works. The position of blind boxes and covered manholes should be well documented. Nevertheless, finding them is often difficult. If they do not contain steel components, a lid with steel bars should be installed on top of the structure in order to facilitate easy location with a metal detector.

Manholes: Inspection chambers or manholes differ from junction boxes with a silt trap in that they provide for ready access if drains require inspection and cleaning. The material can be concrete or masonry, but also redwood has been used successfully. Deep inspection chambers are constructed with a number of reinforced concrete rings. They should be sufficiently large and must be provided with metal rungs fixed in the wall to allow a man to descend to the drain lines. Since the lid of manholes is usually above the soil surface, they are objectionable because of their interference with farming operations. To meet this objection a capped manhole, with the top at least 0.40 m under the soil surface, can be installed with the inconvenience that the top of the manhole has to be dug out for each inspection.

Outlets of Laterals and Collectors

Gravity Outlets

The outlet of laterals and collectors must be protected in case of gravity discharge of the water into an open drain system. The outlet should be reliable since malfunctioning affects the performance of the entire drain or drainage system. The outlet of laterals and smaller collector drains can be protected with a non-perforated rigid pipe made of plastic, coated galvanised steel, reinforced concrete or other materials.

The length of this pipe ranges from 1.5 to 5.0 m, depending on the diameter of the drain pipe, the risk of root penetration from bank vegetation and the danger of erosion under the pipe or at the discharge point. No envelope material (particularly gravel) shall be applied near the outlet and the last few metres of the trench backfill should be well compacted over the entire depth of the trench. The outlet pipe can be connected to, or slid over the drain pipe and at least half of its length should be buried. The main function of drain outlets is to prevent erosion of the ditch bank. For this purpose the unperforated end-pipe must reach far enough out to discharge above the waterlevel in the ditch. Support by a pole or rod may be needed to avoid sagging. Sometimes short non-protruding outlets are used in combination with chutes protecting the side-slope of the ditch.

These chutes can be halved plastic pipes or cement gutters guiding the stream. A non-protruding pipe can also be used where there is danger of ice jams. In spite of many efforts, no adequate solution is yet found to solve the problem of outlet interference with ditch maintenance. Plastic end pipes resist corrosion from chemicals in soil and water but burning off side slopes of ditches as a maintenance measure will be fatal.

Larger collector drains justify the use of a small concrete structure, made of masonry, castin-place concrete, or pre-cast segments. Outlets should be provided with a removable screen to prevent the entry of small animals. Although the outlet into open ditches may be submerged for short periods during storms, they are usually not and should be at least 0.10 to 0.15 m above the water level in the ditch at normal flow.

Pumped Outlets

Pumps are used for the discharge of water from a drainage system into an outlet ditch, when gravity outflow is not possible because of insufficient outlet depth. This situation is common with deep drainage systems that are designed for salinity control in arid and semiarid regions. In other areas they may be needed because of insufficient outlet levels. Collector lines discharge into a storage sump with concrete base, where a float-controlled pump periodically empties the sump.

Pumped outlets are more expensive than gravity outlets, not only because of the initial cost of equipment, but also due to costs associated with maintenance and power consumption. Pumped outlets are equipped with a power unit (either electric motor or diesel engine), and pumps and pipes for lifting collected drainage water to a shallow gravity outlet. Small sumps can be constructed with large diameter plastic, asphalt-coated corrugated steel or concrete pipes while larger sumps shall be made of reinforced concrete rings, masonry or reinforced concrete.

Special Structures

A gradient reducer may be required in sloping lands to reduce excessive flow velocities in drain pipes and prevent erosion and subsequent water movement through channels formed outside the pipe. They can be made of concrete or plastic pipes, or of masonry or concrete. They are in fact blind junction boxes of great height with the entering pipe near the top and the leaving pipe near the bottom of the box.

Although cleaning of properly designed and carefully installed drainpipes should be exception rather than general rule, there may be circumstances where drains require regular cleaning (e.g. if iron ochre is formed). Cleaning of laterals of a composite drainage system, equipped with blind junctions is possible only after dismantling of some of these connections. The provision of special fittings however facilitates cleaning by flushing without having to excavate and dismantle junctions. A concrete tile with steel bars above the access pipe allows easy retrieval with a metal detector from the soil surface.

There can be some reasons to reduce drainage temporarily. Devices for controlled drainage can be installed in open ditches or on subsurface drains. Unperforated pipes with a length of 5 m, leading drains into or from the control structure, should be used to prevent seepage around the structure. Very simple control tools can be used such as an elbow or plug with a riser or a plug with a bypass.

Structures with crest boards are common in open ditches. Very sophisticated structures with crest boards, floats or electric water level sensors in a sump, either located on the drain line or midway between drains, can be used as well. Simple yet reliable control devices can be made locally, however, with available means. Control structures are made of masonry, cast-in-place concrete or pre-cast segments.

Drainpipes serving both drainage and irrigation purposes are sometimes laid without slope. However, this is not necessary as long as the gradient remains sufficiently small. Automatic controls are required to maintain the water level at the drainage outlets, which serve as inlets for subirrigation systems. Subirrigation should not be practised in arid regions where soil salinity is a potential problem.

Envelope Materials

Porous material placed around a subsurface drain, to protect the drain from sedimentation and improve its hydraulic performance, should be referred to as a drain envelope. It is worthwhile to distinguish between the definition and function of an envelope and that of a filter.

During the early development of design criteria for drain envelopes, existing filter criteria were often used as a basis for research. Hence, the word 'filter' is often mistakenly used in reference to drain envelopes. A filter is by definition 'a porous substance through which a gas or liquid is passed to separate out matter in suspension'. Filtration also is defined as 'the restraining of soil or other particles subjected to hydraulic forces while allowing the passage of fluids'. Hence, a filter, used as a drain envelope, would eventually become clogged because particulate matter would be deposited on or in it, reducing its permeability.

Envelopes have the task to improve the permeability around the pipe, and act as permeable constraints to impede entry of damaging quantities of soil particles and soil aggregates into drainpipes. Yet the majority of small particles of soil material and organic matter, suspended in water moving toward a drain, will actually pass through a properly selected and installed drain envelope without causing clogging. The relatively coarse envelope material placed around the drain should stabilise the soil mechanically and hydraulically, but should not act as a filter.

Materials

Mineral envelopes mainly consist of coarse sand, fine gravel and crushed stone, which are placed under and around the drainpipe during installation. If well designed and installed, mineral granular envelopes are quite reliable because they are voluminous and can store comparatively large quantities of soil material without noticeable malfunctioning. As such, they have provided satisfactory long-term service under most circumstances. Traditionally, pit run naturally graded coarse sand or fine gravel containing a minimum of fines is the most common and widely used drain envelope material. Such material can be as permanent as the soil itself. Properly designed graded gravel envelopes fulfil all the mechanical and hydraulic functions of a drain envelope and are the ideal envelope from a physical standpoint.

Graded gravel should be a homogeneous, well-graded mixture of clean sand and gravel free from silt, clay, and organic matter, which could adversely affect its permeability. The use of limestone particles must be avoided, because a high percentage of lime in gravel envelopes is a source of incrustation. In addition, the gradation of a gravel envelope should be made in accordance to prescribed parameters.

The use of gravel as drain envelope has become a bit controversial. One of the conclusions of a symposium held in Wageningen, The Netherlands in 1986 was the following: 'Gravel remains for the time being the most reliable filter material. In view of the cost of gravel the development of design criteria for synthetic materials merits the highest priority'. However, at a conference, held in Lahore, Pakistan in 1990 which was devoted specifically to the design and application of envelopes, it was concluded that engineers who were not familiar with synthetic envelopes, were reluctant to recommend their use. Considering the current tendency, it may be assumed that synthetic envelopes will gradually replace the application of gravel as envelope material in future drainage projects.

Organic Materials

Organic materials, many of which are by-products of agricultural production, have successfully been applied as drain envelopes. They are voluminous, so they can be used in cases where both particle retention and hydraulic function are important. Organic materials may be applied directly on the drainpipe in the trench as loose blinding material, or may be prewrapped around the drainpipe as Prewrapped Loose Materials (PLMs). An intermediate type of application has been in strip-form, applied on top of the drainpipe.

This type of application is now obsolete. Organic envelope materials include chaff, cereal straw, flax straw, rice straw, cedar leaf, bamboo, corncobs, wood chips, reeds, heather bushes, chopped flax, flax stems, grass sod, peat litter and coconut fibre. In northwestern Europe (Belgium, Germany, and The Netherlands), the most common organic envelopes were made from peat litter, flax straw and coconut fibres. The use of fibrous peat litter as a cover layer of drain tiles has been common practice for decades until the end of the 1950s. It was found that the hydraulic conductivity of the peat litter would often decrease drastically due to swelling of the envelope under permanently wet conditions due to e.g. subirrigation.

During the subsequent period, flax straw has been used. It was applied originally as a cover strip and later as prewrapped envelope. The coarseness of the flax envelope did however not always guarantee the particle retention function. On a much smaller scale, other organic envelopes have been applied. These materials were not always available in the required quantities and their handling was often laborious. The use of straw was not successful because it usually decomposed into a low-permeability layer around the pipe.

At the end of the 1960s, coconut fibre was introduced. Being relatively cheap, it soon dominated the market because high quality peat litter became scarce and expensive and because the flax industry declined. Moreover, the finer coconut fibre was considered a more appropriate envelope material than the coarser structured flax straw. Very soon it was discovered that coconut fibres were often subject to microbiological decay. The envelopes were usually fully decomposed after two to five years, particularly if the pH of the soil exceeded the value 6. More than a decade later, many farmers complained about mineral clogging of their drains.

A research project was set up to investigate the problem of mineral clogging. More than 1000 excavations were made and they confirmed that the mineral clogging problems, although partly due to the large effective pore size of the coconut fibre envelope, mainly resulted from the decomposition of the organic substances. In the mid-1980s, various attempts were made to retard or stop the decomposition of organic envelope materials. In Germany and in France a so-called 'Super-Cocos' envelope was introduced.

Its fibres were impregnated with copper sulphate ($CuSO_4$), to kill the bacteria that cause the decomposition.

In addition, some envelopes contained tiny copper wires. 'Super-Cocos' envelopes had limited success because decomposition was postponed for a few years only. In addition, environmental legislation made installation of 'Super-Cocos' illegal in most countries, because the chemical agent leached out rapidly. Coconut fibre envelopes are still being applied in northwest Europe due to their comparatively low price, but their use is declining in favour of synthetic materials.

Organic envelopes have never been popular in countries located in arid climates because the comparatively high soil temperature activates microbiological activity and consequently accelerates their decay. In the irrigated lands of the arid tropics, organic envelope materials usually fail. The successful application of organic envelopes in the Scandinavian countries, where mainly fibrous peat and wood chips were used, was due to the reduced microbiological activity at lower soil temperatures. The service life and suitability of organic materials as envelopes for subsurface drains cannot be predicted with certainty. Eventually, the majority of organic envelopes will decompose, without any serious impact on the structural stability of the surrounding soil.

Hence, these materials should be applied only in soils that become mechanically stable within a few years after installation of the drainage system. In addition, organic envelopes may affect chemical reactions in the abutting soil. This process may result in biochemical clogging of the drain. If iron ochre clogging of drains is likely, reluctance with the application of organic envelopes is justified. Even organic matter that is accidentally mixed with trench backfill material may severely enhance the risk of ochre clogging of the drain. The rapid decay of coconut fibre envelopes has stimulated the search for affordable, synthetic alternatives. The fact that synthetic envelopes can be more easily manufactured according to specific design criteria than organic ones has played a significant role in this development.

Synthetic Envelopes PLM

A synthetic PLM is a permeable structure consisting of loose, randomly oriented yarns, fibres, filaments, grains, granules or beads, surrounding a corrugated drainpipe, and retained in place by appropriate netting and/or twines. Synthetic PLM envelopes are usually wrapped around the corrugated plastic drainpipes by specialised companies and occasionally in pipe manufacturing plants. The finished product must be sufficiently strong to resist handling and installation without damage. Synthetic PLMs include various polymeric materials. Fibres may be made of polyamide (PA), polyester (PETP1), polyethylene (PE), and polypropylene (PP). Loose polystyrene (PS) beads can be wrapped around drains as PLMs in perforated foil or in string netting ('geogrids' or 'geonets').

The beads are subject to compression from soil loads that may reduce envelope permeability. In various European countries where the drain depth ranges from 0.9 to 1.2 m, the effect of the soil load is however relatively small. PLM envelopes made from PP (waste) fibres are increasingly used in northwest. Europe and in arid areas where they replace expensive gravel. Information on some envelope materials.

PLM envelopes made from polypropylene waste fibres (PP- 300) are installed almost exclusively in Belgium for private drainage projects (turnover: 6 percent). PP-450 envelope is a PLM envelope, manufactured from bulk continuous filaments. These filaments are waste when producing woven PP fibre carpets. In The Netherlands, it is by far the most popular envelope material (turnover: 65 percent). PP-700 envelope is a PLM material, made from new PP fibres. Wrapping of pipes with this envelope is comparatively laborious, hence the high price (turnover: 4 percent). It is mainly used for larger pipe diameters (exceeding 160 mm).

Due to the declining availability of PP waste fibres at competitive prices, waste PA fibres are used occasionally.

Contrary to PP fibres, PA fibres absorb water as a result of which the coils may substantially increase in weight. In addition, it is more difficult to process PA fibres to homogeneous prewrapped envelopes because of problems with static electricity. PS-1000 is a PLM envelope material that is manufactured from compressible PS beads in netting and almost exclusively installed in agricultural areas where flower bulbs are grown (turnover: 7 percent). In these areas, the groundwater contains a relatively high amount of suspended particles, and PS-1000 has proven a very reliable envelope. In this application, the higher price of PS- 1000 is a good investment; no farmer can afford to have drainage systems fail.

Synthetic materials deteriorate when exposed to solar (UV) radiation. Experiments with PLM envelopes, made of PP fibres in a temperate climate have indicated that deterioration can be hazardous within three years. The speed of the deterioration will be double in semiarid and arid regions where the average annual radiation is twice that in temperate regions. However, once installed, synthetic PLM envelopes, manufactured from suitable raw material (e.g. recycled PP fibres) are not subject to decomposition. These materials are therefore reliable and affordable substitutes for conventional gravel and organic envelopes. Prewrapping with loose materials is limited to diameters of 200 mm or smaller. Once prewrapped around drains, PLM envelopes have functional properties that are similar to those of geotextiles.

Production of Geotextile Envelopes

According to prEN2 30318 (1998), a geotextile is defined as 'a planar, permeable, polymeric (synthetic or natural) textile material, which may be woven, non-woven or knitted, used in contact with soils and/or other materials in civil engineering for geotechnical applications'. This definition includes application in agriculture since civil engineering incorporates drainage engineering in many countries. Woven geotextiles are manufactured by interlacing, usually at right angles, two or more sets of yarns, fibres, filaments, tapes, or other elements.

Non-woven geotextiles are sheets, webs, or batts, consisting of directionally or randomly oriented fibres, filaments, or other elements. These elements are bonded by mechanical, thermal and/or chemical means. Knitted geotextiles are manufactured by interlooping one or more yarns, fibres, filaments, or other elements. The fibres, used for production of geotextiles are made from the same raw materials as those used for PLMs, namely: polyamide (PA), polyester (PETP), polyethylene (PE), and polypropylene (PP). The fibres of geotextiles may be monofilaments, multifilaments or tapes; the latter either flat, fibrillated or twisted. The combination of raw materials, fibre configuration and weaving, bonding or knitting techniques results in many types of geotextiles which differ widely in appearance, physical, mechanical and hydraulic properties.

In principle, geotextiles may be used as envelope material for drainpipes because they possess two important properties that are required for a drain envelope, namely water permeability and soil particle retention. Moreover, they facilitate the water acceptance of drainpipes, and they convey water in their plane, alongside the pipe wall. Woven geotextiles, however, are seldom used for the manufacturing of drain envelopes. The only justification for this fact must be their comparatively high price, because their specifications are indeed favourable.

In some European countries where organic and synthetic PLMs are used, there is persistent reluctance to use geotextiles as drain envelope because it is argued that their fine texture may enhance mineral and ochre clogging. Yet in countries with a geotextile industry like France, Canada and the United States, geotextile envelopes are applied successfully at a large scale. Laboratory experiments, field trials and practical experiences do not give clear evidence of the clogging risk of properly selected and properly installed fine textured geotextiles. There are, however, circumstances where fine textured geotextiles should preferably not be used.

Wrapping of drains with geotextiles can be done for any diameter. Geotextile strips can be tied around the corrugated drain, or pulled over it after the edges have been sewn together. Geotextiles that are exposed to solar natural weathering are also vulnerable to degradation.

Specification for Drain Envelopes

In 1922, Terzaghi developed 'filter' criteria to control seepage under a dam. These criteria have since been tested for applicability for envelopes around subsurface drains. Terzaghi recommended that the 'filter' material be many times more pervious than the soil base material but that it not be so coarse that the base material would move into the 'filter'. Terzaghi's development has served as a basis for much work done since that time on gravel envelope design. For drain envelopes, his design criteria have been tested and modified, but his original concepts have been generally accepted.

Van Someren reported on the research into and the guidelines for selection and application of drainage materials (pipes and envelopes) in various countries. In Belgium and The Netherlands, efforts were made to develop special design criteria for prewrapped loose materials (PLMs). Conventional design criteria were largely determined by analogue models in laboratories, supported by theoretical considerations, and verified by field trials. Monitoring the flow of water and soil particles near prewrapped drainpipes in the field was not an easy task without disturbing the system. In addition, the data, emerging from field experimentation are inevitably blurred because it is site specific. Results achieved at some places are not necessarily replicable at other locations.

Knops et al. published the first set of comprehensive guidelines for the selection of the then used prewrapped envelopes for use in Dutch soils. Subsequently, a series of research projects and concurrent practical evaluations, carried out by various companies and institutions, have produced design and application criteria for drain envelopes made of PLMs in The Netherlands. Many field surveys have been made into the possible factors that affect pipe sedimentation.

Drain envelopes should meet specifications but visual evaluation of materials is also important. Even if the best materials have been used and all specifications are met, a drainage system will not operate properly if envelopes exhibit some shortcomings due to careless wrapping, handling or installation.

Gravel Envelopes Specifications

Specifications for gravel envelopes are discussed extensively in numerous publications. Sound design criteria for traditional granular envelopes (gravel and coarse sand) are available and have been applied successfully in practice. The US Army Corps of Engineers and the US Bureau of Reclamation have made extensive studies of gravel envelopes. The result is a set of specifications for graded gravel envelopes, which have been successfully used by the Soil Conservation Service, the US Bureau of Reclamation as well as outside the United States.

The gradation curve of a proposed gravel envelope should be matched to the soil to be drained, as well as to the pipe perforations. In addition, gravel should be internally stable to avoid internal envelope erosion. The general procedure for designing a gravel envelope for a given soil is as follows:

- make a mechanical particle size analysis of both the soil and the proposed gravel envelope;
- compare the two particle size distribution curves; and
- decide, by some design criterion, whether the proposed gravel envelope material is suitable.

The involved design criteria consist of rules that prescribe how to derive the particle size distribution, required for a suitable gravel envelope, from particle size distribution data of the soil, in order to guarantee satisfactory service of the envelope.

Terzaghi's criteria

The first criteria, proposed by Terzaghi (US Army Corps of Engineers, 1941) for what he termed a 'filter', are:

— The particle diameter of the 15 percent size of the filter material (D_{15})[3] should be at least four times as large as the diameter of the 15 percent size of the soil material (d_{15}):

$$D_{15} \geq 4d_{15}$$

This requirement would make the filter material roughly more than ten times as permeable as the soil.

— The 15 percent size of the filter material (D_{15}) should not be more than four times as large as the 85 percent size of the soil material (d_{85}):

$$D_{15} \leq 4d_{85}$$

This requirement would prevent the fine soil particles from washing through the filter material. Bertram, Karpoff, and Juusela suggested similar or modified 'filter' design criteria for use with subsurface drains.

Criteria of the US soil conservation service

The SCS (1971) has combined the results of the research on gravel envelopes into a specification for evaluating pit run and artificially graded granular materials for use as drain envelope materials. These specifications are superseded by more recently published specifications, which distinguished between 'filter' and 'envelope'. The recommendation for naturally graded materials or a mixture of medium and coarse sand with fine and medium gravel for use as envelope is:

— $D_{100} \leq 38$ mm.

— $D_{30} \geq 250$ µm.

— $D_5 \geq 75$ µm.

Additional criteria are suggested to prevent excessive fineness of an envelope material, designed to be used for finer textured soils (SCS, 1988):

— $D_{15} < 7\ d_{85}$ but $D_{15} \geq 0.6$ mm.

— $D_{15} > 4\ d_{15}$.

For rigid, unperforated pipes, the US Bureau of Reclamation treats the joint opening, the length of the pipe section, and the hydraulic conductivity of the envelope material as a unified system. Their Drainage Manual contains graphs which consider all these factors.

Since the design of gravel packs for wells is similar to the design of envelopes for subsurface drains, the criteria developed by Kruse for gravel packs may also be used for gravel envelopes. These criteria are based on the ratio of the 50 percent size of the pack (envelope) material to the 50 percent size of the aquifer (soil) and on the uniformity of the textural composition of both the aquifer and the gravel. Kruse observed that sand movement was reduced by decreasing the uniformity of the gravel (i.e. increasing its uniformity coefficient) at all gravel-aquifer ratios and therefore distinguished between uniform soil and gravel pack up to a uniformity coefficient of 1.78 and non-uniform soil and gravel pack for larger values.

The proposed maximum permissible gravel/aquifer particle size ratios for the various combinations of textural composition of both the aquifer and the gravel pack. Besides the 50 percent ratio of filter to aquifer material, Pillsbury also used the standard deviation resulting from the difference between the 95 percent and 50 percent sizes of the grading curve of the gravel envelope divided by 1.645, as a criterion for its effectiveness. Pillsbury presented a graph of the 50 percent size ratio envelope-aquifer vs. this standard deviation which was divided in two zones. Envelopes that fall below the limit line were judged unsatisfactory.

Based on observations of some drain envelopes that had failed in the Imperial Valley of California, Pillsbury recommended an envelope-aquifer ratio of less than 24. He concluded that concrete sand, satisfying the appropriate American Society for Testing and Materials (ASTM) standard with a 50 percent size less than 1 mm and a standard deviation greater than 1.0 would be a satisfactory envelope material under most conditions.

Sherard et al developed filter criteria for protection of hydraulic structures. While not intended for application in subsurface drainage, the principles may equally well be applied for the design of gravel envelopes. The authors established that if a filter did not fail with the initial flow of water, it was probably permanently safe. Well-graded materials were more successful than uniform materials.

Sherard et al reported on tests with fine textured soils and concluded the following with respect to filter and base soil sizes:

— Sandy silts and clays (d_{85} of 0.1 - 0.5 mm) $D_{15}/d_{85} \leq 5$ is safe.

— Fine-grained clays (d_{85} of 0.03 - 0.1 mm) $D_{15} < 0.5$ mm is safe.

— Fine-grained silts of low cohesion (d_{85} of 0.03 - 0.1 mm) $D_{15} < 0.3$ mm is safe.

— Exceptionally fine soils ($d_{85} < 0.02$ mm) $D_{15} < 0.2$ mm or smaller is safe.

Sands and gravely sands containing fine sand fractions and having a D_{15} of 0.5 mm or less would be a suitable filter for even the finest clays. For clays with some sand content ($d_{85} >$ 1.0 mm), a filter with a $D_{15} = 0.5$ mm would satisfy the $D_{15}/d_{85} \leq 5$ criterion. For finer clays, the $D_{15}/ d_{85} \leq 5$ is not satisfied, but the finer soils tend to be structurally stable and are not likely to fail. Finally, Sherard et al. found that well-graded gravely sand was an excellent filter for very uniform silt or fine uniform sand, and that it was not necessary that the grading curve of the envelope be roughly the same shape as the grading curve of the soil. Gravel envelopes that have a D_{15} of 0.3 mm and a $D_{15}/d_{85} \leq 5$ with less than 5 percent of the material finer than 0.074 mm will be satisfactory as envelope materials for most problem soils.

Dieleman and Trafford reviewed criteria for selection of gravel envelope materials and included some comments regarding envelope selection for problematic soils. Dierickx presented a summary of gravel envelope criteria from the United States and the United Kingdom. Prewrapped envelopes may be organic PLM, synthetic PLM and geotextile. Their physical properties such as thickness and mass per unit of surface area are important to check the uniformity of the envelopes, and their conformity with the required design standards. Characteristic opening size, hydraulic conductivity and water repellence determine the hydraulic properties of prewrapped envelopes. When using loose granular materials, particle size distribution parameters may be used as well.

Depending on what kind of drain pipes is used and how envelope materials are wrapped around drainpipes, some mechanical properties of envelopes such as compressibility, abrasion damage, tensile strength and static puncture resistance may be part of the specifications. In The Netherlands, recommendations for the design and application of PLMs have been developed on the basis of concurrent research projects, theoretical studies, mathematical modelling, empirical studies in experimental fields, analogue modelling in laboratories and practical experience covering a 30-year period.

Thickness of prewrapped envelopes

The thickness of prewrapped envelopes serves as a reference for uniformity and conformity. In addition, envelope thickness is found a factor of importance in theoretical analyses as it influences the soil retention capacity, the entrance resistance of drainpipes and the exit gradient at the soilenvelope interface. The main task of an envelope is soil particle retention. In this respect, design criteria for envelope thickness are irrelevant. Thicker envelopes, however, may have higher porosities, which explain their popularity when chemical clogging is anticipated. Therefore, in the envelope selection procedure, envelope thickness is an important parameter, and often significant in terms of safety.

The thickness of an envelope should be a relevant specification if reduction of entrance resistance is envisaged or if reduction of entrance resistance is the only objective to use an envelope. Although a thin envelope may substantially reduce the entrance resistance, the optimal reduction is obtained at a thickness of 5 mm, provided that the hydraulic conductivity of the geotextile is not the limiting factor, which will generally not be the case. When envelopes are used to reduce the exit gradient, the thickness of the envelope is also a relevant design parameter.

The design procedure for envelope thickness, as proposed by Vlotman et al. shows that even thin geotextiles (≤ 1 mm) may considerably reduce the exit gradient at the soilenvelope interface. The larger the diameter of a drain, however, the smaller hydraulic gradients near the drain will be. Hence, 'thick' or 'voluminous' envelopes (i.e. thickness > 5 mm) are generally considered to be safer than thin ones, particularly if the drains are occasionally used for controlled drainage or subirrigation (subsurface infiltration). For PLM, the specification of a minimum thickness was introduced to guarantee a complete cover with a more or less homogeneous envelope. According to the provisional EN-standard, the following minimum thicknesses are required:

— Synthetic, fibrous PLMs: 3 mm (e.g. PP fibres).
— Synthetic, granular PLMs: 8 mm (e.g. polystyrene beads).
— Organic, fibrous PLMs: 4 mm (e.g. coconut fibres).
— Organic, granular PLMs: 8 mm (e.g. wood chips, sawdust).

The provisional EN-standard further specifies that the mean average thickness of each test piece should not deviate by more than 25 percent from that declared by the manufacturer. Geotextiles are available from very thin, sheet-like fabrics to rather thick, mat-like materials.

The mass per unit area is not a selection criterion and therefore not specified. Mass determination can be carried out as a control measure for uniformity and conformity. According to the provisional EN-standard, the mass also may not deviate by more than 25 percent of the mass specified by the manufacturer in order to safeguard a homogeneous product.

Characteristic opening size

The characteristic opening size, derived from the pore size distribution or porometric curve of the envelope, is the most important selection criterion because it determines the effectiveness of the envelope to retain the surrounding soil material. The retention of soil particles is normally not a problem since very fine fabrics are available.

Laboratory research as well as practical experience, however, have revealed that fine envelopes are vulnerable to mineral blocking and clogging. Blocking of an envelope is a decrease of the number of active openings in an envelope that occurs when it is brought in contact with a soil.

Clogging, on the other hand, is a decrease with time of the number of active openings in an envelope due to gradual accumulation of particles inside and on its surface, e.g. by particles suspended in turbid water. Therefore, specifications for envelopes should cover both soil retention criteria and criteria to prevent clogging and blocking of the envelope. Intensive research has resulted in criteria for soil particle retention and in recommendations with respect to the problems of blocking and clogging.

The capability of an envelope to retain the soil material is expressed as a ratio of some characteristic pore size of an envelope to some characteristic particle size of the soil in contact with this envelope. In many countries, the O_{90} is used as the characteristic pore size for organic and synthetic PLMs and geotextiles alike, with a great deal of success. The O_{90} of a drain envelope is the pore size for which 90 percent of the envelope pores are smaller.

The O_{90} value is usually obtained by dry sieving of well-known sand fractions, whereby the envelope itself is installed as a sieve and the retained amount of each fraction is recorded. Wet and hydrodynamic sieving, also applied for this purpose, use graded soil and mostly result in smaller O_{90} values than those obtained with dry sieving.

In 1994, a working group of scientists and engineers in Europe developed a new classification system for PLMs. They introduced three classes of envelopes, depending on the effective opening size of the envelope pores, O_{90}, as follows:

PLM-XF extra fine 100 μm ≤ O90 ≤ 300 μm.

PLM-F fine 300 μm ≤ O90 ≤ 600 μm.

PLM-S standard 600 μm ≤ O90 ≤ 1100 μm.

In the provisional EN-standard only two classes, namely PLMF and PLM-S have been accepted. In The Netherlands, practical guidelines for envelope application consider three 'standard' O_{90} values, namely 450, 700 and 1000 μm, 450 μm being by far the most widely applied, and servicing a great variety of soils. These were accepted after Stuyt (1992a), using field data, confirmed evidence of the soundness of the O_{90} parameter. In Belgium, the O_{90} of a PLM envelope should range between 600 and 1000 μm for official drainage works. A frequently used retention criterion, also called filter criterion or bridging factor of an envelope, is the ratio O_{90}/d_{90}. In this ratio, d_{90} is the particle diameter of the soil in contact with the envelope where 90 percent of the particles, by weight, is smaller.

Numerous other retention criteria have been proposed in the scientific literature, which have been published in comprehensive tables, by e.g. Dierickx and Vlotman et al.. For the design engineer, however, the number of criteria is confusing, the more so because many criteria are contradictory. This fact is self-explanatory, because the criteria were developed under widely different boundary conditions, using many different techniques, equipment and so forth.

Laboratory experiments have unambiguously indicated that the likelihood of soil particle retention is greater when a fabric is thicker. Hence, the characteristic pore size of an envelope may be larger for thicker envelopes, for equal retention. Indeed, retention criteria are linked to envelope thickness. From laboratory studies with analogue soil models, Dierickx, and Dierickx and Van der Sluys derived the following simple retention criteria for subsurface drainage applications:

— $O_{90}/d_{90} \leq 5$ for 'thick' envelopes ≥ 5 mm (PLMs).

— $O_{90}/d_{90} \leq 2.5$ for 'thin' envelopes ≤ 1 mm (geotextiles).

For envelopes with a thickness ranging between 1 and 5 mm, the O90/d90 ratio may be interpolated step-wise or linearly (Vlotman et al., in press).

The step-wise approach gives one value of O_{90}/d_{90} for a range of thicknesses and is somewhat more practical than a linear approach which yields a specific value of O_{90}/d_{90} for each thickness. Retention criteria for thicknesses of PLMs and geotextiles between 1 and 5 mm, according to the step-wise approach are:

— $O_{90}/d_{90} \leq 3$ for thicknesses between 1 and 3 mm.

— $O_{90}/d_{90} \leq 4$ for thicknesses between 3 and 5 mm.

Taking into account the retention criterion of a thin envelope, most problems in subsurface drainage will be prevented by envelopes for which $O_{90} \geq 200$ µm. Field observations of Stuyt confirmed, in a large extent, the laboratory findings. Stuyt investigated the relation between the O_{90} size of envelope materials and the thickness of the sediment layer inside the pipes using a miniature video camera five years after their installation. In total, 9634 m of drains were investigated (184 laterals).

The pipes had outer diameters of 60 and 65 mm. In The Netherlands, sediment layers exceeding 15 mm are generally not tolerated. The d90 size of the soils was approximately 150 µm in most cases. The correlation between the thickness of the sediment layer inside the pipes and the O_{90} size of envelope was significant. Regardless of the O_{90} size, voluminous envelopes retained more soil than thin envelopes. Envelopes with larger O_{90} values, i.e. having larger openings, had poorer soil retention properties.

The raw material from which the envelopes were manufactured was not significant. Stuyt also found that the above-proposed O_{90}/d_{90} ratios were valid for the investigated problem soils. Most of the applied envelopes in the experimental fields had rather high O_{90}/d_{90} ratios.

Experiments with turbid water or water charged with soil suspensions indicate that geotextiles are vulnerable to clogging when $O_{90}/d_{90} \leq 1$. Hence, the ratio $O_{90}/d_{90} = 1$ is the lower limit for soil particle retention, regardless of envelope thickness.

The phenomena of blocking and clogging of an envelope are however not so evident, neither in laboratory experiments with soils, nor in field experiments. Therefore, the lower limit $O_{90}/d_{90} \geq 1$ should be considered a recommendation rather than a rigid design criterion. In the investigation made by Stuyt, envelopes with O_{90}/d_{90} near 1 had such low sedimentation depths that the envelopes appeared to act as filters.

Hence, for thin geotextiles, the O_{90}/d_{90} ratio should preferably be near the upper limit. On the other hand, the upper limit, set to 5 for voluminous envelopes appears safe for voluminous PLMs since a maximum sedimentation depth of 15 mm is tolerated in 60 and 65 mm outer diameter pipes. In soils with some cohesion and, hence, some structural stability, voluminous envelopes with O_{90}/d_{90} ratios as high as 7 have been applied successfully.

In The Netherlands and in Belgium, the successfully applied retention criterion O_{90}/d_{90} for envelopes was therefore adopted as the major design parameter. Recommendations for envelope applications are also based on some additional considerations but the O_{90}/d_{90} criterion is the most important one. Locally made fabrics such as carpet backing, which satisfies or may satisfy the above requirements after some modifications, are equally suitable as imported geotextiles. They may therefore be trusted as envelope materials.

Hydraulic conductivity

The hydraulic conductivity of envelopes should be greater than that of the soil in order to reduce the entrance resistance of drainpipes, so that no hydraulic pressure will develop outside the envelope. From research work of Nieuwenhuis and Wesseling and Dierickx it may be concluded that a substantial reduction in entrance resistance is obtained when

$Ke/Ks \geq 10$, where Ke is the hydraulic conductivity of the envelope and Ks that of the soil. The hydraulic conductivity, perpendicularly to or in the plane of envelope, can hardly be a problem because envelopes are much more permeable than the adjacent soil that they have to retain.

Even under load, the hydraulic conductivity of compressible envelopes will meet the conductivity requirements. If, however, envelopes are brought in contact with soil, particles may fill pores and partly block their openings as a result of which the hydraulic conductivity at the soilenvelope interface will decrease.

In addition, envelopes may clog as a result of particle deposits and/or chemical precipitates, and become less permeable with time. Evaluation of blocking and clogging of envelopes is very difficult. If the lower limit of the retention criteria is taken into account, it may nevertheless be assumed that a favourable hydraulic conductivity ratio is guaranteed.

PLMs do not exhibit wetting problems, yet geotextiles may do and water repellence may be a problem. Water repellence means that a minimum water head is required on top of the geotextile, before water starts to flow through it. Once the water has entered the pipe through the envelope, the repellence problem is solved and will generally not return.

Wettability resistance also decreases when the geotextile is brought into contact with a moist soil. Research work carried out by Dierickx showed that the wetting problem is mainly an initial problem of dry geotextiles.

The initially required head for the majority of the tested geotextiles is smaller than 2 mm. For others, it ranges from 5 to 30 mm; one geotextile required an initial head of 64 mm. Although initial water repellence of envelopes does not seem to be widespread, geotextiles that exhibit this phenomenon should not be used as drain envelope to avoid the risk of soil structure deterioration near the envelope due to the initial stagnation of water. In accordance with the standard on the determination of resistance to water penetration of textile fabrics ISO 811, a testing procedure has been adopted in the countries of the European Union, to examine geotextiles on water repellence in a qualitative manner.

Mechanical properties of envelopes

Mechanical properties of envelopes are mostly of secondary importance. Geotextiles used as drain envelope do not present specific problems since they are designed for, and are normally used in more challenging circumstances. Moreover, problems that develop occasionally because of handling (e.g. tearing) can be repaired before installation. The compressibility of compressible envelopes has a major effect on the characteristic opening size and the hydraulic conductivity.

The opening size normally decreases in compressed state so that a safety factor is built in automatically. The hydraulic conductivity decreases also, yet the highly permeable nature of the envelope ensures that the hydraulic conductivity ratio is met in compressed state. Moreover, the compressibility of coarser envelopes, composed of coarser fibres, is small. Easily compressible thick envelopes, made of fine fibres should not be used as drain envelope.

Abrasion is the wearing of a part of the envelope by rubbing against another material, either during transportation or installation of wrapped drainpipes. Open spots due to abrasion or whatever other cause, noticed before installation, should be repaired in the field, if they are not out of proportion. Abrasion during installation is less likely to occur because of the short time that the wrapped pipe is routed through the machine. Geotextiles are wrapped around drainpipes either manually or mechanically; therefore, a certain tensile strength is required.

Dierickx proposed a tensile strength of 6 kN/m, determined according to the wide-width tensile test. Geotextiles must bridge the corrugations of large drainpipes and may not sag between the corrugations under the soil load. Hence, elongation should be limited, but this requirement is only meaningful if the geotextile is tightly wrapped. Since this has never been a practical problem, elongation requirements have never been put forward.

Resistance to static puncture also is only applicable for drains with large corrugations where a tightly wrapped geotextile bridges the corrugations. The geotextile should withstand the soil load between the corrugations, and puncturing by stones and hard soil clods. These phenomena are simulated by a static puncture test. Through this test, the force required to push a flat plunger through a geotextile can be determined. Since such a problem has never occurred in subsurface drainage so far, no requirements exist.

Cost and Availability of Drainage Materials

Cost and availability of drainage materials are strongly interrelated. Costs vary continuously since these are dependent on various, partly unpredictable factors like currency exchange rates and the cost of manual labour. The cost of gravel envelopes is not specified here because the local availability of suitable granular material is rapidly declining. In addition, the cost of installation is strongly dependent on local circumstances.

In the Integrated Soil and Water Improvement Project (ISAWIP) in Egypt, local gravel envelopes were four times as expensive as imported Canadian synthetic fabric envelopes. In the Fourth Drainage Project of the International Waterlogging and Salinity Research Institute (IWASRI) of Pakistan, the cost of synthetic envelopes was found to be 40 percent lower than that of gravel envelopes.

Installation of synthetic envelopes was easier and faster, too. Thus, even if the price of gravel is competitive, it goes hand in hand with high costs of fuel and manual labour. It is therefore irrelevant to consider the price of the raw material only. Vlotman et al. quote costs of gravel envelopes (material and transport) in various projects in Pakistan. For all projects, the costs of material and shipping of synthetic materials was below the cost of gravel. Unfortunately, the high cost of gravel installation compared to that of installing prewrapped pipes is not included in this analysis. The cost/benefit ratio is certainly in favour of PLM envelopes and geotextiles. PLM envelopes, manufactured from PP fibres and coconut fibres dominate the market in northwestern Europe. PLM envelopes, manufactured from peat fibres are now used only occasionally.

The selection of an envelope material is determined by various factors. The price is obviously important. The ease of handling of the material is also a factor of consideration. Coconut fibre envelopes will release substantial amounts of dust particles during handling and installation, particularly in dry weather; PP fibre wrapping does not.

References

Boussinesq, M.J., "Recherches théoriques sur l'écoulement des nappes d'eau infiltreés dansle sol et sur le débit des sources," *J. Math. Pures Appl.*, 10 : 1–78, 1904.

Boonstra, J., "Estimating peak runoff rates," *In* H.P. Ritzema, ed., *Drainage principles and applications*, 1994, pp. 111–143. 2nd edition, ILRI Publication 16, Wageningen, The Netherlands, ILRI.

Bruggeman, G.A., "Analytical solutions of geohydrological problems," *Dev. Wat. Sci.*, 46: 331, 1999.

Guyon, G., "Considérations sur le drainage des nappes, Théorie et Experimentation," *Bull.Tech. Gén. Rur.*, 1966, 79.

Hunt, R.J., J.F. Walker and D.P. Krabbenhoft, "Characterizing hydrology and the importance of ground-water discharge in natural and constructed wetlands", *Wetlands* 19:458–472, 1999.

Jung, M., T.P. Burt and P.D. Bates, "Toward a conceptual model of floodplain water table response", *Water Resources Research* 40: W12409, 2004.

Moody, W.T., "Nonlinear differential equation of drain spacing," *J. Irrig. Drain. Div. ASCE*, 92(IR2): 1–9, 1967.

Skaggs, R.W., D. Amatya, R.O. Evans and J.E. Parsons, "Characterization and evaluation of proposed hydrologic criteria for wetlands", *Journal of soil and Water conservation* 49:501–510, 1994.

Soil Conservation Service (SCS)., "Hydrology," *National Engineering Handbook Section 4*, Washington, DC, USDA, 1972.

Smedema, L.K., Vlotman, W.F. & Rycroft, D.W., *Modern land drainage. Planning, design and management of agricultural drainage systems,*

Leiden, The Netherlands, A.A. Balkema Publishers, Taylor & Francis, 2004, 446 pp.

Van Drecht, G., *Calculation of steady groundwater flow to ditches*, RID-Mededelingen 83-3, In Dutch with English abstract, The Netherlands, National Institute for Water Supply, 1983.

8

Drainage and Disposal Systems

In any project involving land drainage, it is advisable to work from the downstream part of the system in an upstream direction, that is:

- It is necessary to provide a suitable outlet for the drainage water, either by gravity or by pumping.
- If this outlet alone is insufficient, a main system of open drainage channels and ditches must be constructed to convey the drained water to the outlet.
- If this main system cannot provide adequate control of groundwater levels in the fields, a system of field drains is needed, forming a detailed drainage system, consisting largely of subsurface pipe drains. Open trenches are sometimes used for groundwater control instead, e.g. in drainage of heavy land and in the humid tropics. However, the use of open trenches for subsurface drainage is not generally recommended, as they hamper agricultural operations, reduce the cultivable area and increase the maintenance burden.
- If these field drains are still not able to cope with water stagnating on the surface, additional measures should be taken.
- The division between public main open drainage networks and detailed open subsurface drainage systems is somewhat arbitrary. Ditches serving a few land users are sometimes classed with the former and

sometimes with the latter depending on local circumstances, traditions, and direct responsibility for maintenance. Where any element of the main drainage system is not functioning properly, all upstream facilities cannot fulfil their purpose. Thus, a good outlet and a well-designed and well-maintained main drainage system are prerequisites for adequate field drainage.

Apart from these features dealing with the removal of a certain quantity of drainage water, the quality of the drainage water should also be considered. Within the project area, water quality governs the possibility for its reuse for irrigation, and outside downstream users and downstream ecology may be affected. These quality aspects are becoming increasingly important in drainage projects. Therefore, the layout of the system must minimise its negative environmental impacts.

The FAO Irrigation and Drainage Paper on the management of agricultural drainage water in arid and semi-arid areas considers these factors in detail. Programmes for tree-shaped systems with steady flow, based on the Manning formula, are available at many institutions, waterboards and engineering firms. An example is the HEC-2 programme developed by the United States Army Corps of Engineers. For non-steady-state flow and network structures, more sophisticated methods are needed, such as the programme DUFLOW, based on numerical solutions of the Saint Venant equations, and the HEC-RAS programme, which has been adopted by the United States National Resources Conservation Service.

Various Elements

Where the position and hydraulic characteristics of the outlet are known, the following decisions are of concern in the layout and design of the various elements of the upstream system. The layout will be considered first, followed by a description of the various structures belonging to the main drainage system, such as:

— channels and ditches—they require alongside facilities (tracks, agreements with adjacent land users / landowners) for inspection and maintenance;
— bridges and aqueducts;
— culverts and siphons;
— weirs and drop structures;
— sluices, gates and main pumping stations at the outlet, or any intermediate pumping stations may be considered to belong to this category because they form part of the outlet works.

Attention is needed in order to prevent bank erosion, especially at points where surface runoff collects or field drainage systems are connected with the open channels of the main system.

Drainage Channel Network

The projected main drainage system usually has a branching-tree configuration in which every drop of drainage water has only one way to reach the outlet. However, more complicated network structures are sometimes found, usually remnants from former natural drainage systems. The network depends greatly on the size of the area, its topography, the existing watercourses and the form of its borders.

In a system composed of buried field drains, collector pipe drains, ditches and larger waterways, the length of each successive order determines the distances of the next. Thus, the distances of the first open channels depend on the lengths spanned by the subsurface drainage system. There is a tendency to replace the first open ditches with buried pipes, thus reducing the density of open waterways and consequently saving on maintenance. Another element for the choice of layout is the future maintenance of the main system and its organisation.

The smaller elements can be maintained by a farmer or a local farmers group, be it by hand or by machine. The larger elements can be maintained mechanically by an organisation in which the stakeholders participate and can have indirect influence, e.g. a water board. Within the project area, there may be protected natural reserves. These should be left untouched by the main drainage elements. The channels should keep enough distance from these areas to avoid influencing the underground water currents to and from the reserve area.

Opportunities for improving ecological values sometimes exist in important areas not protected as reserves. Some special drainage with water table management may improve the habitat or ecological values considerably. These potential options should be discussed with stakeholders. Villages and towns and agriculture-based industrial zones in the project area are best provided with a dedicated connection to the public main drainage system to facilitate controlled disposal of polluted water and minimise the risk of improper reuse. Where possible, such urban waters should be treated.

The location of the drainage channel network depends on the topography. In undulating terrain, the watercourses follow the valleys and, thus, the pattern is irregular. However, in flat land, a rectangular layout is usually designed, with exceptions owing to the shape of the project boundary and natural watercourses, or to slight differences in elevation.

Existing waterways are often enlarged, but sometimes they are replaced by a new and wider spaced network of larger channels. These channels should follow the natural drainage paths where possible. As layout and location of elements are highly determined by local circumstances, it is not possible to give more detailed information about these points.

Open Waterways or Channels

Open waterways or channels form the principal part of a system that conveys the outflow from the fields to the outlet. There are two types of layout:

- — a tree structure, where this path is fully determined;
- — a network structure, in which more than one route is available and where the path depends on the local

gradients. The branches of such a network are crossconnected (anastomosis).

Networks, such as the elaborate system of channels of the Mekong Delta, are beyond the scope of this publication as this type is seldom used in new projects. Special calculation methods for flows through networks are available, but they are complicated. In most projects, the tree structure is chosen and a straightforward method of calculation is allowed.

The cross-section of open channels is usually trapezoidal for small drains, and sometimes with a double trapezoid for larger ones. The side slopes are indicated in this publication as the ratio of vertical to horizontal and depend mostly on the type of soil. Some points to consider are:

— Steep slopes save on excavation costs, and such channels occupy less agricultural land, but slopes that are too steep result in bank failures.

— Local experience is the best guide for safe channel side slopes.

— Any slope failures usually occur shortly after construction later, the bank vegetation has a stabilising effect.

— Vegetation (especially submerged plants) obstructs the water flow. Thus, regular maintenance is required. In particular, woody vegetation on the banks must be kept short.

— Lateral groundwater seepage promotes slumping of channel banks. In places with strong seepage, it is necessary to either adopt flatter slopes, provide suitable vegetation cover, or cover the banks with permeable but heavy materials. Geotextiles covered with loose stones are useful in such cases.

— Trapezoidal profiles are designed and built.

However, after many years, they change into more parabolic forms, often with steeper slopes above the usual water levels, covered by vegetation kept short by mowing. In areas with arid climates, the vegetation remains sparse and is confined mainly to the area near and below the water line. The ratio of water depth to bottom width should be kept between certain limits. The calculation of the expected flow rates for the assumed channel dimensions and gradient is based on Manning's formula.

Where the calculated flow rate is too high, the calculation should be repeated with a milder gradient and/or a different y:b ratio. Erosion in water courses should be avoided. At design discharge, the flow velocity must be limited to safe values. At low flows, meandering of the small remaining stream must be prevented as it can undercut the banks. Both can be achieved by placing weirs at appropriate points, so that sufficient water depth is maintained. Another option is to limit the bottom width of the ditch. This is helpful if weed growth is expected to increase with weirs owing to shallow water depths during extended low flow periods.

Parabolic ditch bottoms or small base flow drains in the ditch bottom area have also been used successfully in some cases. Special attention is needed at places where elements of the detailed field drainage systems spill into open waterways. Open drainage channels need regular maintenance. This is because they are susceptible to choking by the growth of aquatic plants and silting up by sediments brought in by uncontrolled surface runoff. In contrast to most irrigation canals, which carry turbid waters, plant growth in open drains is more intensive.

In large channels, a water depth of at least 1 m (1.5 m is better) will hamper the growth of reeds, although submerged and floating plant species may still thrive. In ditches, growth is retarded where they periodically fall dry. A minimum cross-section is often prescribed in order to secure sufficient discharge capacity, especially for small waterways. Such a minimum cross-section has a bottom width of 0.5-1 m and a water depth of 0.30-0.50 m at design discharge, and preferably zero in dry periods. These dimensions can vary according to the machinery available for construction and maintenance.

Although these measures are of some help, periodic maintenance is always needed, especially before the onset of the season in which drainage requirements are highest. Special equipment is available for mechanical cleaning of these open watercourses. This equipment is specific for two maintenance operations: desilting and deweeding. When cleaning an open drain, care should be taken to avoid making the side slopes steeper—so reducing the risk of bank failure.

Bridges and Aqueducts

Where roads and railways cross main waterways, bridges are needed. Irrigation canals usually cross by means of aqueducts. Those that leave the cross-section of the waterway intact have no influence upon the flow in the channel. However, if they are narrower, notably in flat areas, special formulae for flow through openings are used to limit backwater effects. Erosion of the channel under the bridge should then be avoided by not allowing high flow velocities.

Culverts

Culverts are necessary where an open drain crosses a farm entrance or a rural road. One metal, concrete or reinforced plastic pipe buried at least 50 cm deep is commonly used where the water flow is less than 0.5 m^3/s, and two such pipes for discharges up to 1 m^3/s. Their diameter depends on the amount of flow and on their length, but a minimum diameter of 300 mm is often recommended in order to facilitate cleaning. Where the flow is higher, large-diameter pipes, box-type culverts or bridges are used. Calculations for culverts are based on the hydraulics of flow through openings and friction in pipes.

Culverts are usually overdimensioned because they are less able to cope with extraordinary large discharges and to avoid floating debris that may clog them. Whereas open channels may be bank-full or even overflow their surroundings, culverts may be washed out completely and road connections broken. Coarse debris can clog small culverts easily. Hence, they need regular inspection and cleaning. A trash-rack at the entrance can be useful in waterways that carry this type of pollution. A floating beam can hold back floating vegetation. However, both require regular cleaning.

Weirs and Drop Structures

Weirs are used to separate different water levels that would otherwise lead to deep excavations upstream or to an excessive flow velocity and erosion. They can be adjustable or have a fixed crest level. This crest can be sharp or broad, in which case a different coefficient is used for design.

There are various kinds of weir, belonging to two groups:

— *Fixed weirs*: These are the simplest type, but their width may not be ample enough to handle heavy discharges. In this case, "long nose" ("duck bill") weirs may be a solution.

— *Movable weirs*: These are of different types varying from planks or stop logs resting in grove side-walls to self-adjusting valves acting on upstream water levels or forming part of a remotely controlled system.

Drop structures are used in sloping lands where the bottom gradient must be smaller than the ground slope to prevent erosion. They are necessary to maintain the permissible flow velocity, and to dissipate the excess head. Where the energy drop exceeds 1.5 m, inclined drops or chutes should be constructed; and where it is less, straight drop structures are preferred. Weirs and drop structures cause an improvement in water quality through aeration. The overflowing water falls into the downstream section, thereby increasing its oxygen content. As a consequence, degradable organic substances break down more rapidly.

Capacity of Pumping Stations and Sluices

The capacity of pumping stations and sluice structures is determined by the expected flow under unfavourable circumstances, i.e. a design discharge combined with high water levels outside the project. Compared with sluices, pumping stations are not much affected by outside floods, but they are less flexible. Whereas sluices increase their capacity at high internal levels, the capacity of pumps is almost fixed. A pumping station is needed where the outer water level is either always or for long periods above the desired inner water level.

Drainage by pumping is often only necessary in the rainy season; then a sluice or a gate is combined with the pumping station to allow discharge at low outside levels. A pumping station consists of: an approach channel, which enables uniform flow; a sump, where drainage water collects; a suction pipe; a pump or group of pumps; and delivery pipes with outlets to the receiving waterbody. Generally, the peak flow is discharged through several pumps and the base flow through only one. However, other combinations may be used according to the circumstances.

An additional standby pump is usually included for safety reasons and to enable repair of one of the other pumps without losing design capacity. Three types of pumps are commonly used according to the drainage flow to be elevated and the lift. Where the flow is less than 200 litres/s even if the lift is high, radial pumps are recommended because they have greater flexibility in relation to flow variations. Axial pumps are most suitable for water flows of up to 1 m3/s and low lift (2-4 m). Where the outer level is almost constant, and the water transports vegetation or other debris, an Archimedes screw is appropriate.

The choice of pump type also depends on local conditions such as availability, experience, maintenance possibility, and prices. These pumps are driven by diesel engines or by electricity where enough power is available on the spot. Where the electricity supply is sufficiently reliable, notably in periods when rainstorms occur, electrically driven pumps are preferable to diesel pumps because they require less attention, maintenance and management, and they can be automated easily. The design of pumps and pumping stations is discussed in Wijdieks and Bos.

Where pumping stations are essential in a drainage network, then they require competent operation and solid preventive maintenance routines. For sluices, the tides or high outside levels may hamper the drainage discharge. In combination with the storage possibilities inside, this leads to solutions that are highly dependent on local conditions. These require special calculations, based on: numerical estimation of the storage and the levels inside; the water levels outside; and the flow through the sluice opening when the inside levels are higher.

Field Drainage Systems

Structures to conduct this surface and subsurface water safely into the main drainage system without causing erosion are necessary. These usually consist of rigid pipe sections for subsurface drain outlets as well as for many surface water exit points. Pipe outlets should discharge above the water level of the receiving channel in order to facilitate visual inspection and to prevent bank erosion.

An alternative is protection of the bank below the pipe outlet by chutes lined with grass, concrete or rock. For large outlets, special structures are made. Their dimensions and construction materials depend on local circumstances and on the amount of water to be discharged. Attention must be given to the effect of these drainage connection structures on mechanical cleaning operations, especially where they occur at many points. In this respect, chutes and special constructions are more convenient than protruding pipes.

Design Requirements and Criteria

Capacity Requirements

The capacity requirement of the main drainage outlet system is that it maintains sufficiently low water levels under unfavourable conditions. This means that in wet periods occurring with a frequency of once in 5-10 years, it must provide an adequate outlet for the field drainage systems so that these outlets still have free discharge into the main system or—if that is not always and everywhere possible—are only submerged temporarily and slightly.

The water levels are governed by the following data:

- — specific discharge (drainage coefficient);
- — design discharges of channels;
- — hydraulic gradients and geometry of channels;
- — head differences for culverts, bridges, weirs, sluices and pumps.

For the design, the channel system is divided into sections in parts that are small enough to be considered homogeneous in discharge and gradient, so that within each section the bottom width and water depth will be the same. Bridges, culverts, weirs, sluices and pumps are treated as separate structures.

Specific Discharge and Drainage Processes

Main drainage system discharges are generated by various field drainage processes. Of these, the surface drainage processes are usually the most critical. The specific discharge is the rate at which excess water must be removed by the system without difficulty. In humid climates, it is the runoff that occurs on average from rainfall with a frequency of once in 5-10 years, increased with water from other sources. It is usually expressed in millimetres per day and is later converted into a drainage coefficient expressed in litres per second per hectare for further calculation.

A less probable precipitation event is sometimes taken in order to check the safety of the system and the extension and duration of the flooding under extreme circumstances. Gumbel's method may be used to predict such rare phenomena from a limited amount of data. For humid areas, there are several methods to estimate the discharge intensity for design. However, the effects of climate change must be kept in mind. Under arid conditions, not more than 1.5-2 mm/d is usually required for salt control and irrigation losses. However, where rainy seasons occur, this coefficient may be much higher.

In principle, expected seepage should be added. However, in humid areas, it is usually negligible compared with the rainfall. However, in arid regions, the drainage coefficient is low and seepage can be of comparable magnitude or even higher. Moreover, where seepage is saline, then the soil salt balance of the rootzone may be affected significantly in arid regions. Impermeable surfaces, such as areas with bare rocks, asphalt road, buildings and horticulture under glass or plastic, have a large specific discharge. This is because infiltration in the soil is impossible.

In agricultural areas, the influence of these areas is usually of minor importance. However, built-up and covered areas tend to become more extensive over time, especially where large cities come into existence or where covered horticulture or orchards with intensive surface drainage systems become widespread. Problems may arise in periods with exceptionally intense rainfall. Where such problems occur, they may lead to a revision of the drainage system in a distant future. This involves mainly an increase in the open water storage by allowing certain low-lying areas to be fiooded in critical periods as compensation for loss of soil and land surface water storage.

Design Discharges of Channels

In areas with rainfall, all discharges from upstream sources must be added in order to calculate the necessary transport capacity at the end of each section below the stated highest admissible water levels as well as for any other construction belonging to the drainage system, taking into consideration the effects of retention and different travelling times in the discharging subdrainage catchments. The design flow for return periods of 5-10 years is determined at representative places:

— at the outlet of contributing smaller channels;

— at the beginning and end of each of channel section;

— at other constructions;

— at the final outlet.

At these places and also in channel sections in between, control points are located where characteristics such as surface elevation and other data are measured. In reality, the flow rate changes with time, and storage in the channels will cause the discharge process to be nonsteady. Nevertheless, in many cases, the channel storage is relatively small in comparison with the storage in a pre-wetted soil. Therefore, the storage leading to nonsteady effects is mainly located in the subsurface drainage system of the fields and not so much in the main system.

Moreover, the channels are often short enough to ignore outflow retardation by travelling discharge waves, so that steadystate calculation is often a good approximation. However, in cases where surface runoff is important, the storage in the fields is much smaller and, consequently, the design discharges for the main system become far higher. Non-steady-state calculations for runoff normally begin at the upper end of open drains and proceed downstream.

Determinations on the time of the runoff peak, its shape and duration are used to calculate the size of outlet drains. Generally, the empirical method of the unit hydrograph is applied. In this case, the shape of each contributing area and the slope of the watersheds enter into the channel-sizing equation. For steadystate calculations in a short channel section, the flow is taken as flow from upstream sections plus the inflow into that proper section. Both flows are calculated as the product of the specific discharge (q in litres per second per hectare) multiplied by the contributing area (A in hectares) in order to obtain the flow (Q in cubic metres per second).

This gives a slight overestimation, but the difference is on the safe side. A reduction is often applied to the upstream flow from large areas (Q). It is in the form of an exponent n (< 1). Where the local rainfall is patchy, the area for considering reduction is above some 1 000 ha. Where the local rainfall is widespread, reduction may be applied if the upstream area is larger than 50 000 ha. Recommended area reduction factors (n) can be consulted in Smedema, Vlotman and Rycroft.

In irrigated areas of arid regions where rainfall is negligible, the accumulation of discharges from different parts of the system is not necessary unless flooded rice is grown. In this case, a high drainage discharge capacity is required at the end of the growing season. This is because all farmers want to evacuate the remnants of the standing water layer in a relatively short time. As not all fields are irrigated at the same time in non-rice areas, the peak discharges from the different sections do not occur at the same time. Thus, the peak discharge from the entire system is less than the accumulated peak discharges from the sections. FAO has provided values for the multiplying factors to determine the design discharge for collector drains as related to the fraction of the area that is irrigated simultaneously.

Exceptional Discharge

In order to check the safety of the system during discharges with return periods of 50- 100 years (which may occur at any time), the calculations are sometimes repeated with a value of q of 1.5-2 times higher than the design coefficient. For such rare occasions, the water levels may become much higher than normally allowed, but disasters such as serious floods or severe damage must be avoided.

Hydraulic Gradients

The hydraulic gradient of a channel section is the slope of the hydraulic energy line along the channel. At low velocities, this line is almost equal to the slope of the channel water surface. It must be more or less parallel to the slope of the land along the channel. Initially, the average hydraulic gradient available for gravity discharge can be chosen to be approximately the same as the surface gradient. Where the terrain is completely flat, it is necessary to choose a small hydraulic gradient.

This must be enough to allow sufficient water flow. However, considering the need to avoid erosion, the flow velocity should not exceed 0.5 m/s. In silty soils, it may be as low as 0.20 m/s. The bottom slope of an open-channel section should normally be equal to the average hydraulic gradient. Values of 0.05-0.1 per thousand are common in flat areas. To create a higher gradient in these cases, discharge by pumping from one section into another could be considered. However, the capitalised operation costs may easily exceed the saving on channel dimensions.

Longitudinal Profile of a Channel

In a longitudinal profile of a channel, the level of the strip of land along the top of the banks and the water levels to be tolerated at design discharge should be indicated, together with the location of buildings and confluences. Sudden changes in the gradients should be avoided and, where necessary, occur only at the limits of a section. Where sudden water level changes are required by the topography, weirs are needed. Their location follows from land surface measurements. Head losses caused by weirs and other structures must be shown in the channel hydraulic profiles. Weirs and culverts cause differences in head between their upstream and downstream ends, and weirs in particular lead to backwater effects that may be noticeable far upstream in flat country.

Calculation Programmes for Drainage Design

Since the advent of the electronic computer, models have found wide application. For drainage, various models are used in research and engineering. Universities and research institutes have developed sophisticated models, and governmental institutes, engineering companies and individual consultants use various calculation methods for design.

Information on applications of GIS for planning and design of land drainage systems can be consulted in Chieng. Computer programmes for drawings, such as topographic and layout maps, and detailed design of open drains and ancillary structures of the main drainage system are widely used by engineering firms. Additional information on computer applications related to land drainage is given in Smedema, Vlotman and Rycroft.

The CD-ROM version of this FAO Irrigation and Drainage Paper includes several programmes for drainage design, largely based on formulae given earlier in this publication. The aim is not to clarify the underlying fundamentals or provide great sophistication, but rather to facilitate their direct application to drainage design under practical circumstances. In addition, some related problems are addressed that have influenced the design itself, such as backwater effects and seepage.

The programmes are in FORTRAN and run under both Microsoft Windows and DOS. Inputs are in the form of questions and answers. Choices between various possibilities have to be made by typing certain numbers, and input data have to be provided in the same way. The units are metric, in accordance with FAO standards.

General Structure

The programmes have a common basic structure, allowing easy retrieval. For this purpose, certain rules have to be followed regarding notation of decimals, the abbreviated name of the project and the location.

The following items are considered:

- The programme mentions its name and purpose in order to check that it is appropriate. If not, it can be terminated easily.
- A point must be used as decimal separator. A question is raised about national usage; if a comma is the norm, a warning is given.
- A "project" name of a maximum of four characters is required (letters or numbers in single quotes). This shortness is because of the restricted length of filenames under DOS.
- Within this project, several locations can be used, each of which characterised by a name of a maximum of ten characters in single quotes (letters or numbers).
- At the end of the session, the project receives a unique name for the output file, showing the results for the various locations.

— For easy retrieval, all filenames are listed in a file LIST, where indicates the kind of programme used.

Specific Programmes

Extreme values

Extreme values are the largest and smallest elements of a group. In many cases, they obey Gumbel's probability distribution. Applications are: the highest precipitation in a certain month and the highest discharge of a river in a year. The programme GUMBEL allows an easy method for interpolation and extrapolation. For a given return time, it calculates the value to be expected.

A poor fit indicates uncertainty in the basic data; a distinct upward trend that the data do not obey the GUMBEL distribution and that the extrapolated values are far too low. In this case, other methods must be used. By extrapolation, a prediction can be given for return periods of 100 or 1 000 years. However, the uncertainty becomes considerable at such long times. Nevertheless, such extrapolation is valuable for engineering purposes, such as for the height of river embankments needed to withstand a "100-year" flood.

The flood will almost certainly not take place after 100 years, but it has a probability of 1 percent of occurring next year (and maybe tomorrow) and has a good chance of occurring in a lifetime. Last, it must be borne in mind that natural and human-induced changes may influence the events in question.

Examples are: the increase in impermeable surfaces (roads and cities) and deforestation will increase drainage flows; and climate changes (whether natural or human-induced) will have either positive or negative effects. For drainage design, return periods of 2-10 years are often taken (2-5 years for agricultural field systems, 5-10 years for the main system), but these must be far higher if human safety is involved. For example, in the Netherlands, return periods up to 10 000 years are used for sea dykes in critical areas.

Measuring Soil Permeability

The auger-hole method is widely used for measuring soil permeability. The water level in an auger hole is measured before pumping, and afterwards its rise is determined. In dry soils, the fall of the water level after filling can be observed, but this "inverse" method is less reliable. Moreover, some soils swell slowly and have a much lower permeability in the wet season than when measured dry.

The programme AUGHOLE can process the data obtained for both the normal and inverse methods. The results within the same auger hole are usually quite consistent. Where more than one observation is made in the same hole, the programme takes the average and gives its standard error. When large variations are encountered, a message appears: "Not reliable".

Between different holes, even nearby ones, differences may be considerable owing to local soil variations. However, in predicting drain spacings, these errors are diminished because the resulting spacings are proportional to the square root of K. The resulting K values can be used as input for programmes such as SPACING and the NS series.

Piezom

In an open auger hole, a kind of average permeability is measured for the layers between the groundwater level and the bottom of the hole. Where data are required for a specific layer, Kirkham's piezometer method can be used. The auger hole is covered by a tightly fitting pipe, and, with a narrpwer auger, a short open cavity is made below its open bottom. Alternatively, an auger hole is covered partially by the open pipe and the remainder forms the cavity below. In the former case, the diameters of pipe and cavity are different; in the latter, they are almost equal. As with the auger-hole method, water levels are measured at different times. The permeability is measured of the layer in which the cavity is located.

Spacing

The programme allows the calculation of spacing of pipe

drains under steadystate conditions in cases where upward or downward seepage towards deeper layers is insignificant. If such seepage is considerable, ARTES must be used instead. If nonsteady situations have to be considered, a preliminary steadystate solution by SPACING can be checked with programmes from the NS series. In SPACING, up to five soil layers can be considered, and anisotropy may be accounted for. However, in practical cases, sufficient data are seldom available and estimations are usually needed. Nonetheless, the effect of additional layers and anisotropy can be investigated by entering trial values.

NSABOVE, NSDEPTH and NSHEAD

These programmes analyse the nonsteady behaviour of a proposed or existing drainage system after complete or nearly complete saturation of the soil after heavy rainfall, snowmelt or irrigation.

NSABOVE can be used if the drains are at the impermeable base, so that the flow is above drain level only. The programme gives the expected lowering of the groundwater table from zero to a given depth within a given time. These data can be based on agricultural requirements that depend on the tolerance of the crop or on soil tillage and trafficability needs.

NSDEPTH is used if also deeper layers take part in the drainage process. As in NSABOVE, the criterion is the lowering of the groundwater. It uses numerical calculations, and allows inclusion of the radial and entrance resistances near the drainpipe and the limited outflow capacity of the drainpipe and the main drainage system.

NSHEAD is similar to NSDEPTH but mentions the head above drain level instead of the water depth.

Artesian Conditions

Artesian conditions may cause upward seepage where a deeper lying aquifer is under pressure, or natural drainage where the pressure is lower than the pressure of the shallow groundwater. These conditions can exert a large influence on the layout of a subsurface drainage system. Strong upward seepage can lead to failure, whereas natural drainage can diminish the required intensity and even make subsurface drainage unnecessary.

In principle, geological information and a model such as SAHYSMOD are needed. However, for a first estimate, ARTES can be used to see whether serious effects are to be expected. At this stage, good data about the aquifer and the top layer are seldom available, but estimates can provide some insight about the effects to be expected. The programme gives two solutions—one for a wet and one for a dry season. The latter is usually critical because of capillary rise and salinisation hazards.

Wells

Instead of drainage by a network of pipes or open channels, a network of wells may be used (vertical drainage). However, this method can only be used under specific circumstances:

- — A good aquifer must be present.
- — This aquifer must have sufficient contact with the overlying soil, so that pumping can influence the groundwater levels.
- — There must be no danger of attracting brackish or saltwater from elsewhere.
- — Overpumping must be avoided, although it may be allowed temporarily.

Under favourable circumstances, such a network may be useful. The programme provides a simple approach for steadystate conditions. However, a more sophisticated method, based on geohydrological studies, is recommended for estimating the effects such as overpumping and salinisation.

Drain Diameters

DRSINGLE and DRMULTI

For long drains and wide spacings, and especially for

collectors, it is often more economical to start with a small diameter and change to a larger size further on. Moreover, different materials may be used in the same drain. The programme DRMULTI calculates such "multi" drains. Which of the two programmes should be chosen depends on the local availability of pipes and on local prices.

Main Drainage System

BACKWAT

Where the main system discharges into a river or the sea, or indeed any waterbody that shows fluctuations in water level, backwater effects occur. Especially during high outside levels, they interfere with the discharge from above. Open outlets may even allow a rapid flooding of the area.

The programme gives an initial steadystate approach to such backwater effects. It gives the steady backwater curves, positive at high outside levels, negative at low ones.

Interceptor Drains

INCEP and INCEP2

In undulating terrain, waterlogging and salinisation often occur at the foot of slopes or below higher irrigated or rainfed lands. Stagnation of groundwater also occurs in places where the thickness of an aquifer or its permeability diminishes suddenly. This may be caused by the presence of a rock sill. A related problem is the interception of water leaking from irrigation canals. The programmes calculate the width of a drain trench or ditch sufficient to cope with the intercepted flow. INCEP is valid for a homogeneous profile, INCEP2 for a drain or ditch located in less permeable topsoil.

The size of the drains needed to discharge the flow must be found from the programme DRMULTI, using the inflow per metre given by the programmes INCEP. In homogeneous soil, a normal drain trench is wide enough in many cases. However, drains in a less permeable top layer require much wider trenches or broad ditches. A practical solution is to put more than one drain in such locations. As the hydrological circumstances are often complicated and little known, the programmes can only give global guidelines. In practice, the problem is usually solved by trial and error—if a single drain is insufficient, more are added.

References

Boumans, J.H., "Stromingsformules voor drainage ribbelbuis," *Cultuurtech. Tijdschr.*, 26: 41–46, 1986.

FAO., *Materials for subsurface land drainage systems*, by L.C.P.M. Stuyt, W. Diericks & J. Martínez Beltrán, FAO Irrigation and Drainage Paper No. 60 Rev. 1. Rome, 2005, 183 pp.

Irwin, R.W., "Corrugated pipe flow rate from pipe geometry," *J. Irr. Dr. Eng.*, 1984, 237–241.

Van der Beken, A, De Troch, F., De Somer, M. & Zuidema, F.C., "The lateral inflow into submerged drains," *Bull. Int. Ass. Hydrol. Sci.*, 17: 273–289, 1972.

Zuidema, F.C. & Scholten J., "De afvoercapaciteit van geribbelde plastieken draineerbuizen," *Van Zee tot Land*, 49: 73–102, 1972.

9

Surface and Subsurface Drainage

SURFACE DRAINAGE SYSTEMS

On flat agricultural lands, with slopes often below 0.5 percent, ponds form where the infiltration into the soil is less than the amount of water accumulated after rainfall, snowmelt, irrigation or runoff from higher adjacent places. In cold climates, a combination of snowmelt and frozen subsoil is particularly troublesome, while in dry regions so is an irrigation followed by unexpected heavy rain. Ponds form on the ground surface, especially where the infiltration rate is below the precipitation intensity. This process also occurs where the groundwater is deep.

Fine-textured soils, especially ones with a weak structure, and soils that form crusts easily are most susceptible to low infiltration and ponding. The cause is usually at or very near to the ground surface, in the form of natural pans or human-induced compacted layers such as plough soles. Deeper layers of low permeability are sometimes the cause of the formation of a perched water table. Another cause of pond formation is insufficient subsurface drainage, causing groundwater tables near or even above the surface.

In this case, the flow is not restricted by insufficient infiltration into the soil but by the limited discharge of groundwater. The two processes sometimes interfere. A temporary high groundwater level may cause slaking and crust formation, which then causes stagnation of water on the surface, even after slight rains. Such pools tend to become larger during further rains. In temperate climates with low-intensity rainfall, the precipitation rate is usually lower than the infiltration into the soil. Thus, surface runoff is limited to special cases, i.e. steep and barren slopes, very impermeable soils, land compacted by heavy machinery during the harvest of root crops, and soils that are susceptible to crust formation.

In summer, the land is dry enough to absorb even a heavy shower. In such climates, subsurface discharge dominates. In places where surface runoff occurs, local or temporal solutions are common. In climates where rainfall is more torrential, the volumes of surface runoff can be considerable, especially on soils with low infiltration rates and from land that has been conditioned to reduce the incidence of ponding in high-value vegetable crops and orchards. Both rainfall intensity and infiltration rate are functions of time, and their combination leads to a critical period when conditions are worst. Such a period usually lasts a few hours.

Where the type of agriculture requires its removal, as is usual in flat areas, surface drainage is needed. In addition, part of the infiltrated water must also be removed by subsurface drainage, but this flow comes later. Surface water stagnation has negative effects on agricultural productivity because oxygen deficiency and excessive carbon dioxide levels in the rootzone hamper germination and nutrient uptake, thereby reducing or eliminating crop yields. In addition, in temperate climates, wet places have a relatively low soil temperature in spring, which delays the start of the growing season and has a negative impact on crop yields. Excess water in the top soil layer also affects its workability.

The length of the critical period of crop inundation must be determined from local experience as it varies according to climate, soils, crop tolerance, crop development stage and cropping conditions. In humid temperate regions, common field crops, such as maize and potato, usually require designs to remove ponded rainfall from the drainage area within 24 hours. Some higher value horticultural crops may require a 6-

12-hour removal time during the growing season, while other crops can tolerate ponding for a couple of days.

The objective of surface drainage is to improve crop growth conditions by providing timely removal of excess water remaining at or near the ground surface before the crops are damaged. Surface drainage is also needed to guarantee soil workability and trafficability, so preventing delays in soil preparation operations and harvesting, respectively. In order to do this effectively, the land surface should be made reasonably smooth by eliminating minor differences in elevation. It should preferably have some slope towards collection points, such as open field drains or shallow grassed waterways, from where water is discharged through outlets especially designed to prevent erosion of the ditch banks.

Land smoothing is the cheapest surface drainage practice and it can be performed on an annual basis after completion of tillage operations. On sloping and undulating lands, generally with natural slopes of more than 2 percent, ponding is not usually much of a problem, except for a few small depressions. However, the resulting runoff may cause severe erosion during heavy rains. Where this occurs, reshaping of the land surface into graded terraces that generally follow the contours is needed in order to promote the infiltration and the storage of useful moisture in the soil.

The necessary earth movement can at the same time be used to fill the small depressions where runoff tends to collect. Earth movement is expensive (at least US$2/m^3 even in low-income countries) and it requires considerable expertise and further maintenance because of soil subsidence and settling. The field surface drains (furrows or shallow ditches) discharge into a network of open ditches or grassed waterways and larger watercourses. The main drainage system removes excess water to points outside the project area. Care must be taken to protect stretches where surface runoff collects and enters into field surface drains or where these drains enter larger ones. These are the points where gullies can start and where sediments enter into the main drainage system. At these transition points, provisions are needed to control erosion, even in flat lands.

Flat Lands

In flat lands, the approaches to cope with excess surface water depend on the circumstances. Where high groundwater is not a problem, surface systems, such as furrows and raised beds, are sufficient. However, a system of shallow ditches, combined with surface drains where necessary, is often used to cope with high groundwater as well as surface water.

Downstream End of a Field

Where there is a small slope, surface runoff from an individual field may be discharged into a furrow running parallel to the collector ditch at the downstream end of the field. Bank erosion may be prevented by a small dyke along the ditch. The water collected in the furrow is then discharged safely into the open ditch through a short underground pipe. The same drainage outlet is generally used for removing excess irrigation water, especially in rice fields.

Ridges and Furrows

Where crops are grown on ridges with furrows in between, their somewhat higher elevation protects plants from inundation. The furrows also serve as conduits for the flow of excess water, which is collected by an additional furrow at the downstream sides of the field and discharged into the ditch in a similar way as described above.

The ridged fields may have a small slope towards the sides. Where fields are made highest in the middle, this position can also be used for irrigation supply to the furrow. The length and slope of the furrows depend on the field dimensions and the soil conditions. The length usually ranges from 150 to 250 m. The slope along their length is usually some 0.5-5 per thousand. This guarantees a flow velocity of less than 0.5 m/s, low enough to prevent erosion on most soils.

Convex Raised Beds

In flat lands with low infiltration rates, surface runoff is facilitated by shaping the land into raised beds with a convex form between two furrows. Beds run in the direction of the prevailing slope. A rather low lateral direction slope of these beds is sufficient. In some soils, beds that are too high may become subject to erosion. Raised beds can be made on-farm by repeated directional ploughing or by land grading.

The intervening furrows are shallow enough to be passable for agricultural implements and cattle. These furrows should have a slight longitudinal slope for their discharge, either directly to the collector ditch, as in grassland where the soil is sufficiently protected, or to a system with a downstream furrow acting as a surface drain.

While normal ploughing operations must always be carried out in the same way the beds were ploughed originally, all other farming operations can be carried out in either direction. The beds have a length of about 100-300 m. The bed widths and their slopes depend on soil permeability, land use and farm equipment. Some recommendations, according to Raadsma and Schulze and Ochs and Bishay, are:

- — 8-12 m for land with very slow internal drainage (K = 0.05 m/d);
- — 15-17 m for land with slow internal drainage (K = 0.05-0.10 m/d);
- — 20-30 m for land with fair internal drainage (K = 0.1-0.2 m/d).

The elevation of the beds, i.e. the distance between the bottom of a furrow and the top of the bed, can range from about 20 cm for cropland up to 40 cm for grassland, where land covering reduces erosion hazard. The furrows between the beds are normally about 25 cm deep with gradients of at least 0.1 percent. The bedding system does not provide satisfactory surface drainage where crops are grown on ridges, as these prevent overland flow to the furrows. Bedding for drainage is recommended for pasture, hay or any crop that allows the surface of the beds to be smoothed. It is less expensive but not as effective as a parallel furrow drainage system. The system cannot be combined with surface irrigation, although sprinkler and drip irrigation remain possible.

Parallel Field Drainage Systems

Parallel field drainage systems are the most common and generally the most effective design recommended for surface drainage of flat lands, particularly where field surface gradients are present or constructed. Parallel field drainage systems facilitate mechanised farming operations. Shallow field drains are generally parallel but not necessarily equidistant, and spacing can be adjusted to fit farm equipment. The spacing of parallel field drains depends on the crops to be grown, soil texture and permeability, topography and the land slope.

Drain spacing generally ranges from 100 to 200 m on relatively flat land, and it depends on whether the land slopes in one direction or in both directions after grading. Parallel field drains should usually have side slopes not steeper than 1:8 and longitudinal grades ranging from 0.1 to 0.3 percent. To enable good surface drainage, crop rows should be planted in a direction that will permit smooth and continuous surface water flow to the field drains. Ploughing is carried out parallel to the drains, and all other operations are perpendicular to the drains.

The rows lead directly into the drains, and should have a slope of 0.1- 0.2 percent. Where soil erosion is not probable, the row slope may be as high as 0.5 percent. Under some conditions, deeper field drains are also used to provide subsurface drainage. In several places, especially at the outlets, small filled sections with culverts are often needed to provide access to the fields.

Parallel Small Ditches

This system employs small ditches 0.6-1.0 m deep. It is used with the dual purpose of removing surface runoff and controlling high water tables. The system is especially useful where the groundwater stagnates on a poorly permeable layer at shallow depth, but also functions to prevent a high rise of the real groundwater during wet periods. In this case, all farming operations are carried out parallel to the drains. The distance between the small ditches is usually 50-100 m, with a length up to 500 m. The length of these ditches depends on the spacing of the ditches receiving the discharge.

Longitudinal slopes of 2-5 per thousand are recommended in order to secure their discharge and, at the same time, to prevent their erosion. Where surface runoff is a problem, shaping the land will provide either one- or two-sided discharge to these ditches. Erosion protection for parallel ditches is sometimes needed, especially on arable land. In pastures, the side slopes of the ditches are usually covered with vegetation, and protection against surface runoff is seldom needed.

Sloping and Undulating Lands

With undulating and sloping lands, there is ample opportunity for free surface runoff, and often also for natural underground drainage to a deep water table. However, erosion of such lands often causes sedimentation elsewhere, while the runoff leads to inundations in the lowest parts of the area. Groundwater flow may cause seepage in lower places.

Sloping Lands

Where surface runoff threatens agricultural fields in sloping lands, small cross-slope ditches can be made at their lower end, running almost along contours. Ditch spacings depend on factors such as gradient, rainfall, infiltration into the soil, hydraulic conductivity, erosion risk and agricultural practices. No general rules can be given. Surface runoff is discharged into open collector ditches running in the direction of the natural slope to discharge into a main waterway. The open collector ditches should not erode. Therefore, the slope of the land should be not more than a few percent; otherwise, the collector ditches must be provided with weirs or drop structures.

To facilitate agricultural operations, the ditches can be made passable for machinery or provided at their ends with a dam and an underground pipe leading to the collector drain. The width of the dam and the length of pipe depend on the type of machines to be used, but a pipe length of about 5 m is sufficient. When constructed, the excavated materials should be used in low areas and on the downhill side of the ditches.

Random Drains

Random drains are applicable where fields have scattered isolated depressions that cannot be easily filled or graded using landforming practices. The system involves connecting one depression to another with field drains, and conveying the collected drainage waters to suitable outlets.

Drain depths should be at least 0.25 m, with dimensions depending on the topography of the area and on discharge design, considering the contributing area. This minimum depth is usually applied in the uppermost depression areas. To permit crossing by farm machinery the side slopes should be no steeper than 1:8. The spoil or excavated material from random field drains should be used to fill small depressions or be spread uniformly so that it does not interfere with surface water flows. Smoothing is sometimes required in order to improve the effectiveness of the surface drainage in some of the flatter parts of these fields.

Undulating Lands

On undulating lands, the layout of an improved drainage system must follow as much as possible the natural topography of the existing watercourses. In narrow valleys, one open drain is usually sufficient, but wider plains may require interceptor or diversion drains, often in addition to contour embankments at the foot of the surrounding hills, to protect areas from flooding caused by surface runoff from higher lying adjacent lands. Infiltrated water can reappear in the valley as seepage, causing a more permanent type of waterlogging, and in dry climates severe salinisation. This situation is common near the foot of hills bordering flat valleys, and also in low-lying lands that receive tail-end water and/or seepage water from adjacent higher lying irrigated areas.

The type of interceptor drains used depends on the relative amounts of runoff and seepage. The former usually dominates, in which case open ditches are needed. Their side slopes, especially the upstream one, must be very flat in order to prevent erosion, and grassed waterways are often useful. A grassed filter strip is also recommended for the upslope side of the interceptor ditch. It catches sediments carried by the water and prevents erosion of the slope.

Where seepage is of importance, deeper ditches are required, and pipe drains can be used if there is little or no surface runoff. Some narrow valleys still have a considerable longitudinal slope, the open ditch being liable to erosion. By grading the land, the valley may be divided into compartments separated by small transverse dams. An open drain situated near the centre of the valley collects water from upstream and transports it to the lower end of each compartment. There, a weir or drop structure leads to the next one. In some cases, pipes can be used in combination with inlets of surface water situated at the downhill end of the compartment. Such inlets can be made from largediameter plastic pipes surrounded with gravel.

Cross-sections of Ditches

Ditches must have enough capacity to transport the drainage water in wet periods. However, they are sometimes made wider than needed in order to create more storage in the open water system. Such temporary storage is a good way of diminishing the peak outflows from the area, as occurs after heavy rains. Thus, it reduces the required capacity of downstream constructions, such as the larger watercourses, culverts, and pumping stations.

The cross-sections of ditches are usually trapezoidal although small ones may be V-shaped. Their dimensions vary according to: the expected runoff, the necessity for open water storage, the capacity to be passable for machinery, the risks of bank erosion, and the available means for maintenance. Because ditches tend to hamper agricultural operations, passable drains are often used, designed with respect to agricultural land use rather than on hydraulic properties. Where they tend to erode, they are sown with grasses (grassed waterways). However, grassed waterways are not always a solution because sometimes the grass does not grow or it does not survive the dry season.

Ridges and furrows are made by ploughing with ridge-forming agricultural machinery, passable ditches usually by grader, and steeper ones may be constructed by a special plough that shapes the required profile in one pass. Larger ditches are usually made using a backhoe.

Discharge of Excess Surface Water

The discharge of excess surface water to be expected determines not only the dimensions of the structures described in the previous sections, but also those of drainage elements of the main system further downstream. Peak discharges are caused almost exclusively by rainfall or snowmelt; in rare cases, they stem from irrigation losses. First, the drainage coefficient, defined as the rate of water removal per unit of area, is estimated. Then, the flow rate, which varies with the size of the area, is calculated. In flat lands, design discharges depend on the amount of excess rainfall to be removed by the surface drainage system during the critical period.

The first item can be estimated from the water balance or through empirical formulae. In sloping land, although surface stagnation is generally not the problem, design discharges are needed to dimension the different components of the main drainage system. Discharges stem from overland runoff processes in the basin considered. There are several methods to obtain the hydrograph of the basin; some of them are quite sophisticated. Therefore, before describing some of the methods for calculating design discharges in flat and sloping areas, the following section considers some principles on surface runoff.

Concepts of Overland Runoff

Water Balance

The amount of excess rainfall to be drained superficially during a critical period can be estimated from the water balance at the ground surface:

$$Sr = P - E - I_{nf}$$

where:

E = direct evaporation (mm);

I_{nf} = infiltration into the soil (mm);

P = total precipitation (mm);

Sr = excess of water at the soil surface (mm).

The excess of surface water is generally drained freely in sloping lands, but commonly through surface drainage systems in flat lands (Ds). Part of the infiltrated water sometimes interflows through the topsoil (Di), but most replenishes the unsaturated zone and percolates, recharging the groundwater table (R). Where natural drainage is not sufficient, subsurface drainage (Dr) is required (figure 1).

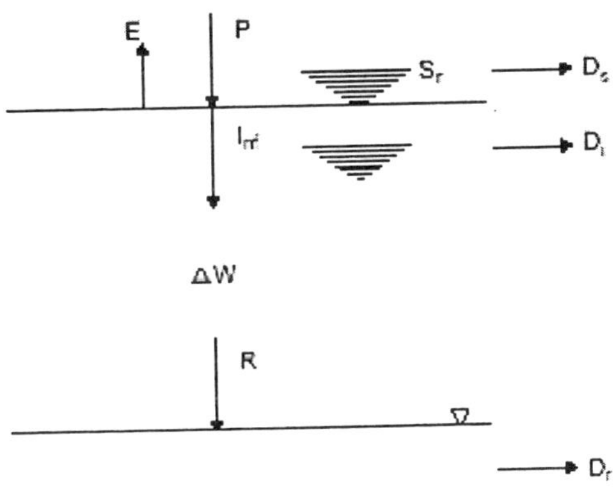

Figure 1. Components of the water balance after a heavy rain

The evaporation in a period of a few hours is usually small and negligible compared with the other terms of the water balance. The amount of rain to be expected with a given frequency in a critical period can be estimated from meteorological data. For extreme values, Gumbel's method may be used to obtain such forecasts. Generally, only rainfall data for 24 hours are available. However, the length of critical periods can be 6-12 hours and, moreover, heavy rainfalls usually occur in this time interval. Nevertheless, estimations for these short periods can be made, for example with the following coefficients:

$$P_6 / P_{24} = 0.5\text{-}0.7$$

$$P_{12} / P_{24} = 0.6\text{-}0.8$$

where:

P_6 = estimation of the amount of rainfall in 6 hours (mm);

P_{12} = estimation of the amount of rainfall in 12 hours (mm);

P_{24} = amount of precipitation in 24 hours (mm).

Where only rainfall data for one-year return period are available, estimations for 5 and 10 years can also be made with the following coefficients:

$$P_{T5} / P_{T1} = 1.5\text{-}2.0$$

$$P_{T10} / P_{T1} = 1.7\text{-}2.5$$

where:

P_{T1} = precipitation for 1-year return period (mm);

P_{T5} = precipitation for 5-year return period (mm);

P_{T10} = precipitation for 10-year return period (mm).

For snowmelt combined with a still frozen soil, it should be expected that the total precipitation accumulated as snow during the foregoing frost period (minus some evaporation by sublimation) will become runoff within a few days. More important to a water balance is the infiltration, which depends greatly on the soil properties. While coarse sands will take almost any rainfall intensity, finer sands can show surface runoff during heavy showers.

Silt loams have a tendency to form crusts, and some clay soils have a low infiltration rate whereas other well-structured ones may remain very permeable. However, all soils show an infiltration rate that varies with time. When still dry at the surface, they have a much higher intake rate than after wetting. The main reason is that at the beginning the hydraulic gradient between the wet top and the dry subsoil is very large.

Eventually, the intake rate becomes constant because the soil is ultimately saturated and the hydraulic gradient has become unity owing to the effect of gravity only. Another cause of reduced infiltration is that clay swells on wetting. The determination of infiltration forms the main difficulty, but field methods are available.

Hydrographs of Surface Runoff

In an agricultural area, surface runoff depends on some physical characteristics of the basin, such as its form and size, soil conditions, land slope, natural vegetation and land use. The peak flow of drainage water also depends on the characteristics of the main drainage system, such as drain density, cross-sections and gradients of the watercourses, as well as their maintenance conditions.

After a certain amount of precipitation (P), the specific discharge of surface drainage water at the outlet of the basin (q) increases progressively during the elevation time or time to peak (t_e). Once the maximum value (q_M) is reached, the specific discharge decreases progressively during the recession timc (t_r). The time interval between the average time of the storm (t) and the time when maximum discharge occurs is called the lag time (t_d).

The hydrograph for total drainage discharge can be obtained by superimposing the groundwater hydrograph on this hydrograph. In a basin, the values of the times described above are constants they depend on the concentration time (t_c). This is the time interval since the beginning of the storm and the moment when runoff coming from the most distant point from the outlet of the basin contributes to the water flow at the outlet.

For basins less than 1 500 ha, the concentration time can be considered equal to the time to peak. If the duration of the storm is less than the tc, only part of the basin contributes to the peak flow at the outlet; if the is higher, the whole area contributes, but generally the rainfall intensity decreases with time. The tc value depends on the flow velocity and on the length of each section of the main drainage system:

$$t_c = \sum_{i=1}^{n} \frac{l_i}{V_i}$$

where:

t_c = concentration time (s);

l_i = length of section i of the main drainage system (m);

v_i = average flow velocity in section i (m/s).

Where the drain hydraulic cross-section, the slope and the Manning coefficient are known, the flow velocity in the watercourses can be calculated with the Manning formula. The flow velocity on the ground surface depends on the covering (land use) and slope. However, in agricultural areas of less than 50 ha, the concentration time can be estimated with the empirical formula developed by Kirpich:

$$t_c = \frac{K^{0.770}}{3080}$$

where:

h = difference in elevation between the most distant point in the basin and the outlet (m);

l = maximum distance between the above two extreme points (m);

$s = h/l$ = gradient;

$K = \frac{1}{\sqrt{3}}$ = basin constant (m).

Determine Design Discharges

Different methods have been developed to determine peak water flows and design discharges. The approaches differ from sloping lands, where surface runoff is free, to flat lands. In addition to this distinction, the selection of the appropriate method for a specific project area depends on data availability.

Batch Method for Flat Lands

For humid flat lands, a simple and approximate method, called the batch method, is based on rainfall, outflow, and storage in different reservoirs, this being:

— soil storage;

— storage in channels and ponds;

— storage by field inundations.

In the batch method, a water balance is set up in order to obtain an approximation of the consequences of different drainage coefficients on crop growth during the critical period.

Empirical Formulae for Flat Areas

In flat areas, empirical formulae can also be used. Special formulae are available for specific regions and their use is recommended if they are based on sufficient experience. As an example, the Cypress Creek formula, developed for flat lands in the east of the United States of America. As actual conditions may differ in a project area, this formula can only be used as a first approximation to be verified later.

Statistical Analysis

The maximum discharge at the outlet of the main drainage system can be determined statistically where a data series of measured flows is available covering a period of at least 15-20 years in an area where the hydrological conditions and the land use have not changed during the historical period considered.

Unit Hydrograph

In agricultural areas, long data series of measured flows are rarely available to determine statistically the design discharge. However, in basins of 10 000-50 000 ha, where it is possible to assume that 2-6-hour storms are covering the area uniformly, flows have sometimes been measured for different duration rainfalls. Therefore, some hydrographs are available. By using these hydrographs, a precipitation/surface runoff relationship can be obtained. This can be used to predict the surface runoff for other series of rainfall data. The unit hydrograph developed by Sherman is based on this principle.

Rational Formula

In agricultural areas of 100-200 ha, surface runoff is produced just after precipitation where the storage capacity of water in the soil is low. No unit hydrographs are usually available, but there are sometimes some gauge points in the main drainage system. In this case, with water flow data and the characteristics of the section affected by the measurement of the water flow, a relationship can be established between the amount of surface runoff and rainfall. This relationship can be applied to other areas with similar characteristics to the reference section.

Curve Number Method

In agricultural areas, the most frequent case is to have rainfall data available but no surface runoff information. In this case, surface runoff can be estimated with the available rainfall data and information on the physical characteristics of the basin concerning the rainfall/runoff relationship, by using a method based on this relationship. A method widely applied is the Curve Number (CN) method. This method was developed by the Soil Conservation Service (SCS) after studies and investigations made in basins with surface area below 800 ha.

To apply the CN method three phases are followed:

— The amount of surface runoff expected after the design rainfall is estimated, by considering the physical characteristics of the basin.

— The distribution of the estimated runoff during the storm period is determined by using an undimensional hydrograph.
— The maximum value of the specific discharge is determined in the hydrograph obtained for the total discharge. Then, with the surface area value, the peak flow at the outlet of the main watercourse draining the basin is calculated.

This method has a wider scope of application than the rational method as it can be applied in basins with a surface area of several thousand hectares. However, the result obtained can only be considered an estimation of the peak flow. This must be further checked with measured flows in gauge stations in similar locations to the place of application (as original curve numbers were developed in the United States of America).

Subsurface Drainage Systems

In flat lands, subsurface drainage systems are installed to control the general groundwater level in order to achieve water table levels and salt balances favourable for crop growth. Subsurface drainage may be achieved by means of a system of parallel drains or by pumping water from wells. The first method is usually known as horizontal subsurface drainage although the drains are generally laid with some slope. The second is called vertical drainage.

A system of parallel drains sometimes consists of deep open trenches. However, more often, the field drains are buried perforated pipes and, in some cases, subsurface collector drains for further transport of the drain effluent to open water are also buried pipes. The drainage water is further conveyed through the main drains towards the drainage outlet. Less common are vertical drainage systems consisting of pumped wells that penetrate into an underlying aquifer. In sloping lands, the aim of subsurface drainage is usually to intercept seepage flows from higher places where this is easier than correcting the excess water problem at the places where waterlogging occurs from shallow seepage.

Singular and Composite Drainage Systems

There are several options for the layout of systems of parallel drains:

— singular drainage systems consisting of deep open trenches flowing directly into open outlet drains of the main system;

— singular drainage systems consisting of perforated pipe field drains (laterals) flowing directly into open drains of the main outlet system;

— composite drainage systems in which perforated pipes are used as laterals and closed or sometimes perforated pipes as collector drains. The latter discharge into the main drain outlet system.

As open trenches hamper agricultural operations and take up valuable land, field drainage systems with buried perforated pipes are usually preferred. Several factors must be considered in order to select the appropriate drainage system, such as:

— the need to discharge surface runoff;

— the slope of the land to be drained;

— the depth of the lateral outlets;

— the maintenance requirements and possibilities;

— the design depth of the water table.

Singular subsurface drainage systems, with pipe laterals only, are appropriate:

— where, in addition to the subsurface flow, it is necessary to discharge excess rainfall through a shallow surface drainage system;

— where a certain amount of water must be stored in the open drains in order to reduce the peak flow in the outlet system;

— in very flat lands where the drainage flow is high and the available slope is low.

Composite subsurface drainage systems, with pipe lateral and collector drains, are generally recommended in the irrigated lands of arid regions because:

- — The depth of field drains is usually greater than in the temperate zones and, consequently, large excavations would be required if open ditches were used as field or collector drains.
- — The excess rainfall is generally negligible; as a consequence, drainage rates are low and thus the discharge of a considerable number of parallel pipe drains can readily be collected and transported by a subsurface collector system.
- — Weed proliferation increases the maintenance costs of open ditches.

This type of system is common in the Nile Delta, Egypt, where subsurface drainage systems discharge only the necessary leaching to control soil salinity and keep the groundwater level sufficiently deep to prevent salinisation caused by capillary rise of saline groundwater. Composite systems are also recommended in: sloping areas where soil erosion must be controlled and/ or drainage problems are mainly manifest in patches or in topographic lows; in areas where the land is very valuable; and in the case of unstable subsoils that cause unstable banks of open drains.

In some areas, especially where the maintenance or availability of deep open drains is difficult, groups of pipe collector drains discharge into tanks (sumps), from where the water is pumped into a shallow main outlet system. This is the case for arable crops and mango orchards in some parts of the Lower Indus Plain, Pakistan, and in some areas of the Ebro Delta, Spain, where horticultural crops are grown.

Controlled drainage is sometimes used to slow drainage during dry periods, and increasingly to control water requirements of rice in rotation with dry-foot crops. Then, the water table is maintained at a higher level by technical means, such as temporary plugs in subsurface drainage systems, raising seasonally the open drain water levels, or rising lateral/collector pipe outlets. Thus, a certain amount of water is saved from flowing away during droughts, or when fields are flooded during a rice crop.

In Egypt, during rice cultivation in otherwise dry-foot crop cultivated land, such plugs are used to close the orifice in the bottom part of a specially constructed overflow wall inside inspection maintenance hatches of composite drainage systems. Water tables can also be controlled by subirrigation, where water from outside sources flows into the drain if the outside water level in the whole area is kept high for a considerable period. Apart from these uses, it is effective for preventing clogging with iron compounds, and the outflow of nitrates from the drainage system may be reduced by denitrification. However, great care should be taken with such systems in arid areas subject to salinisation. Although there are no physical restrictions on the length of subsurface field drains, it is usually governed by the size of the agricultural fields and the maintenance requirements of the drain.

In composite systems, the same applies to the length of collectors. Where cleaning is required, the maximum length of pipes is usually limited by the maximum length of the cleaning equipment, which is about 300 m. However, where there is enough slope and no constraints on designing pipe drains longer than 300 m, extended systems can be designed. However, they require a suitable access construction for cleaning devices at about every 300 m. As longer drains require larger diameter pipes, maintenance hatches should be installed to facilitate the connection between pipes of different diameters, as well as for inspection and cleaning, notably in the case of collector drains. Accessible junction boxes should be placed at the junctions between laterals and collectors.

Designing Subsurface Drainage Systems

In designing horizontal subsurface drainage systems, in addition to the drain length B described above, the following dimensions are needed:

— drain depth Z;

— drain spacing L;

- drain slope s, or total allowed head loss in the drain at design discharge intensity H;
- drain diameter d.

Moreover, for composite systems, the dimensions of the collectors must be determined. The type of pipes and possible types of protective drain envelopes must be selected, preferably from among the types and sizes that are readily available in the country. In addition, the method of installation (trenchless or in dugout trenches) and the method of maintenance must be chosen. The design dimensions, such as the average drain depth, drain slope and allowed head losses, are usually the same for large areas, often over an entire project. Sometimes, they are prescribed quantities.

On the other hand, drain spacings, lengths and pipe diameters may vary considerably from place to place, as spacings depend on crops and soil conditions, lengths on the system layout, and diameters on spacings, lengths and slope. The lengths and diameters of field and collector drains depend considerably on the dimensions of the plots to be drained, thus on the parcelling of the area. Both are interrelated, as the longer the drains are, so the greater their diameter must be.

As the price of pipes increases with diameter, in the case of long drains, where all diameters of pipes are readily available, it can often be profitable to begin upstream with smaller pipes, using increasing diameters further downstream. The switch in diameter has to be done at a logical place (maintenance hatch), otherwise mistakes can be made during installation and/or problems may occur with the cleaning of the drains. The drain spacing is also related to cost. In singular drainage systems, the costs are almost inversely proportional to the spacing.

The drain spacing and the drain depth are mutually interrelated - the deeper the level of the drains so the wider the drain spacing can be. Thus, increased spacing might lower the amount of the subsurface drainage work, and consequently the costs. However, in some cases, the cost advantages of greater drain depth may be offset by an increase in construction cost per unit length, by larger diameter of field drains, by higher costs of deeper collectors and ditches, and by costlier O&M, especially where deeper drains need a lower outlet level.

Moreover, deeper drainage is often restricted by other factors, e.g.: by soil conditions, as in heavy clay soils with shallow impervious layers; outside water levels, as happens in lowlands; or, less frequently, by the availability of appropriate machinery. For example, in Egypt, during often relatively short fallow periods, groundwater must be lowered in order to limit topsoil salinisation by capillary rise. Detailed cost calculations resulted in the conclusion that deeper and wider spaced drainpipe installation only entailed modest installation cost savings owing to the extra cost stemming from larger drain diameters. For example, a system where the water level between drains is designed at 1.50 m below field level with a hydraulic head of 0.30 m requires a drain depth of 1.80 m and drain spacing of 80 m.

During the fallow period in this arid area, the actual water level between drainpipes will be slightly higher than 1.80 m. Where the pipes are installed at 1.60 m depth to fulfil the requirement of a water table at 1.50 m, the pipes have to be spaced at 50 m. This means a depth gain of only 20 cm, for a cost increase of about 60 percent. During the fallow period, the water table depth is then about 1.60 m (instead of 1.80 m). However, in the heavy clay soils of the Nile Delta, capillary rise is very slow, and as irrigated cropping intensity is high, both depths are sufficient to prevent soil re-salinisation.

Once a design drain depth has been selected, there are two different approaches to calculating the drain spacings:

— for conditions of steadystate groundwater flow towards the drains, where the flow in wet periods is assumed to be constant in time;

— for non-steady-state flow conditions, where flow is time-dependent.

In the former case, an outflow intensity, which is assumed constant, is used as a criterion; in the latter case, the time to obtain a given drawdown after a critical recharge event is taken as design datum. The steadystate method can be used where the recharge to the water table is approximately constant during a critical period. Then, it is possible to design the system with a discharge equal to the recharge. If, at a design water table height, the inflow of water to the soil is constant and equal to the drain outflow, the water balance in the saturated zone is in equilibrium and the groundwater level remains at a constant depth.

In practice, steadystate flow is a good approximation:

— in temperate zones with long periods of low-intensity rainfall that are critical for drainage;

— in areas recharged by deep upward seepage from a semi-confined aquifer;

— in areas where there is lateral seepage from outside waterbodies;

— in irrigated lands where water is continuously applied through high-frequency irrigation methods, such as drip irrigation and central-pivot systems.

The steadystate approach is less applicable where high recharges occur in a short period of time only, such as after heavy irrigation or sudden rainfall. In this case, the water balance is not in equilibrium as when the recharge is higher than the discharge, the groundwater level rises; and when the recharge ceases, the system is still draining, and the water level falls. The conditions where soil water storage is important in design are frequent in:

— areas with heavy showers of short duration, common in some Mediterranean areas and in the humid tropics;

— in irrigated lands with intermittent irrigation where applications of 60-120 mm are common.

However, under certain assumptions, nonsteady drainage flow conditions can be converted mathematically to steady flow conditions. Therefore, steady flow considerations can be used as a substitute for processes that are essentially nonsteady in nature.

Drainage Requirements

In humid temperate areas, agricultural drainage must be able to prevent damage to crops in periods with abundant rainfalls occurring with a frequency of once in 2-5 years. In arid areas, drainage should prevent the accumulation of harmful amounts of salt and provide adequate drainage after a heavy irrigation or after heavy rains as occur in monsoon-type climates. Artesian conditions often lead to upward seepage flow of water from deeper layers. This flow has a great influence on the design of a drainage system. It often makes a "normal" drainage unable to prevent waterlogging or salinisation.

Thus, extra measures are necessary in upward seepage areas. Where the seepage water can be reused, vertical drainage may be an option for controlling the water table. Drainage requirements result in two important factors for drainage design, which are used in the steadystate determination of drain spacing: the specific discharge q; and the hydraulic head midway between two drains h, which should be available for causing the required groundwater flow. This head represents the drain depth Z minus the required groundwater depth z.

For nonsteady calculations, an additional input parameter is needed. This is the storage coefficient. Therefore, the dimensions of a subsurface system depend on the following drainage criteria: ÿ the design groundwater depth z or the depth of the water table below the soil surface, midway between drains, during times of design discharge;

- the outflow intensity q or the design discharge of the drains per unit area, and usually expressed in millimetres per day;
- in nonsteady cases, the time in which the groundwater should regress from the initial high water tables to a given water table depth is used. This recession time depends on crop and temperature; for horticultural crops, it is usually short, especially under high temperatures.

Fundamental criteria such as design groundwater depth and design outflow are derived from guidelines, local experience, research plots, theoretical considerations and models. For example, the DRAINMOD model allows evaluation of criteria or checks on those derived by other means. The following sections provide some indications for values of these drainage criteria.

Design Groundwater Depth

Critical to crop growth and soil trafficability is the depth at which the groundwater remains/fluctuates under critical circumstances. At design discharge for field crops, this depth z is usually of the order of 0.9 m, but it varies by crop, soil and climate. For shallow-rooting horticultural crops on pervious soils, depths of 0.5 m may be reasonable. Tree crops require greater depths than vegetables, but the latter can stand water near the surface only for a few hours and, thus, are vulnerable to extreme high water table situations, especially when temperatures are high.

In temperate zones, controlled drainage permits two design groundwater depths: a deep depth to provide aeration and trafficability in periods with excess of water; and a shallower depth to facilitate subirrigation in dry periods. Controlled drainage also permits high water levels for nitrate reduction and preventing iron precipitation in the pipes. In climates with low-intensity rainfall, the following minimum depths to the steadystate design groundwater depth during short wet periods are usually recommended:

- 0.3-0.5 m for grassland and field crops for design outflows of about 7-10 mm/ day;
- 0.5-0.6 m for vegetables grown on sandy loam soils.

In arid areas, two design depths are frequently required: one during the cropping season to provide aeration to the rootzone (unless rice is grown); and a second one for fallow periods in order to prevent capillary rise and associated salinisation (where seepage from irrigation elsewhere would cause too high groundwater levels). As the drainage discharge is also different during the cropping and the fallow seasons, the drain spacing/depth has to be designed for the most critical period (the smallest h/q), bearing in mind the required groundwater depth during the fallow period (smaller h, lower q). In irrigated lands, the following design depths for groundwater for steadystate design outflow can be used as a starting point:

— 0.8-0.9 m for field crops;

— 1.0-1.2 m for fruit trees, depending on soil texture.

In the case of irrigation of rice, controlled drainage permits the elevation of the groundwater level up to the ground surface in order to prevent excessive water losses. Here, there is no danger of salinisation owing to the absence of upward flow in the inundated soil. To control capillary rise and related soil re-salinisation processes, groundwater must remain below a certain depth in periods without rain or irrigation.

This safe design depth is determined mainly by the capillary properties of the soil and the salinity of the top layers of the groundwater mound. In particular, silts and silt loams require deep drainage. Where the critical depth to control capillary rise is excessive and higher groundwater levels have to be accepted, then, in order to secure acceptable soil salinity levels, the salts accumulated during the fallow period must be leached by irrigation where there is no excess rainfall.

Design Discharge

In humid temperate areas, the design discharge occurring with a frequency of once in 2-5 years is usually taken as the design criterion. Under these circumstances, crops should not suffer from waterlogging. In arid climates, prevention of salinisation is the main purpose of drainage, and for most cases a discharge capacity of 2-4 mm/d is sufficient for leaching. In humid tropical areas (including those with monsoon climates), the rains are often so heavy that the infiltration capacity limits recharge, and surface runoff may occur.

In addition, the subsurface drainage system is usually unable to cope with the inflow. The same applies to other climates with intense rains. In such cases, a combination with a surface drainage system is needed. After the rains, when the soil is saturated, the subsurface drainage system then lowers the groundwater to a sufficient depth in a reasonable time, whereas in the dry season it prevents the accumulation of salts.

The exact figures for the design discharge q are extremely dependent on the local climate conditions and/ or irrigation practices. Therefore, the outflow intensity is usually derived from local experience. Where considerable seepage occurs, the amount of seepage water must be added to the design discharge, and the pipe sizes adjusted accordingly. For example, this is the case where relief wells are used to tap the aquifer - the drains must be able to convey this extra amount of water.

Groundwater Lowering

In the non-steady-state design method, both z and h are functions of time. After a heavy rain or irrigation, the groundwater should fall a given depth in a given time so that its depth z increases. Because Z cannot change with time, h also falls by the same amount. Such a requirement can be used to calculate drain spacings. In this nonsteady case, the storage coefficient and not the discharge is used as an input parameter. In this case, the drain discharge rate varies with time.

Where heavy rains or irrigation have caused water to stand on the surface, the following criteria for the lowering of the groundwater could be used under nonsteady flow:

— for horticulture, a lowering after complete inundation of 0.30 m in 4-6 hours;

— for most crops in hot climates, a lowering or 0.30 m in 1 day;

— in cool climates, a lowering of 0.20 m in 1 day.

In irrigated lands, in addition to these criteria, the soil provides storage for the percolation water, and the drainage system must be able to remove this storage before the next irrigation. Therefore, between two irrigation applications, the drawdown of the water table must be similar to the elevation produced by the irrigation water losses. A low outflow criterion is usually sufficient for this purpose. For example, where 40 mm of percolation is stored in a soil with a storage capacity of 5 percent, this gives a rise of 0.8 m.

The groundwater level must be low before the following irrigation, for example 30 days, and the stored water must be removed in this period, requiring on average a drainage coefficient of 1.3 mm/d. Under these circumstances, the best approach is to design the system with a steadystate method and a low outflow intensity, and then to simulate its behaviour after complete flooding. If the outcome is not satisfactory, the steadystate discharge must be changed by increasing the steady outflow criterion. This will lead to a narrower spacing, which can be tested again for its nonsteady behaviour. The process is repeated until a satisfactory solution is obtained.

System Parameters

Drain Depth

The selection of the drain depth is a crucial and early decision in a drainage project. This is because of the technical aspects involved, and because of the direct influence of the drain depth on the overall cost of the system. As mentioned above, deeper drains allow wider drain spacings with fewer drains per unit area, but other factors, such as construction and O&M costs of field and main drains and outlet structures, play a role in the overall cost.

The depth of the laterals Z is equal to the sum of the depth to the water table z and the hydraulic head h both taken midway between two drains. Under steadystate conditions, the required groundwater depth must be adjusted by the head loss h required to cause groundwater flow towards the drains:

$$Z = z + h$$
$$h = Z - z$$

where:

h = head loss for flow in soil, at design discharge (m);

z = groundwater depth midway, at design discharge (m);

Z = drain depth (m).

As mentioned above, the design value for z depends on climate, crop requirements (crop calendar, rooting depth, crop salt tolerance), and soil and hydrological conditions. Moreover, to select an adequate drain depth Z, the hydraulic conductivity and the soil stability of the layers situated above the impervious barrier should be considered. Unstable soils such as quicksand are to be avoided. Although quicksand can be handled, it requires a special installation technique with sometimes modified machines.

In addition, the drain depth is often limited in practice by the water level at the outlet of laterals or collectors into the main drainage system. The minimum depth of open trenches for subsurface drainage is about 0.6 m, and for pipes it is about 0.8 m. Pipes installed at a shallower depth may become clogged if crop or tree roots penetrate into the drain through the pipe slots. In addition, shallow pipe drains can be damaged during subsoiling operations, which are common in the management of clay soils with low permeability. In cold climates, pipes must be deep enough to prevent freezing.

Drain Spacing

Drain spacing is an important factor because the cost of subsurface drainage is related closely to the installed length of drains per unit area:

$$C = \frac{10000}{L} C_u$$

where:

C = installation cost of the system (in terms of monetary units per hectare);

C_u = cost per unit length of installed drains (in terms of monetary units per metre);

L = drain spacing (m).

Although the field drains form a major component of the cost, collectors and the main drainage system are important items, as are the capitalised costs of O&M. Therefore, if it is decided to install deeper drains to allow wider spacings, the additional costs of the required deeper main system must be compared with the savings on field drains. There are various methods of calculating drain spacings from the drainage requirements and the soil characteristics. Of these, the soil permeability, the layering and anisotropy are especially important factors.

The calculation methods fall into the two categories mentioned above: steady and nonsteady flow. In steadystate calculations, the inputs are the design head loss h or midpoint water table height and the design discharge q. In the nonsteady case, the design factors comprise a prescribed increase in groundwater depth z with time in combination with the storage coefficient. Steadystate methods may form a first step in designing drain spacing, but nonsteady methods can represent the changing conditions more accurately.

Therefore, as a second step, drain systems designed tentatively with steady criteria may be subjected to more realistic, variable inputs in order to evaluate the design. In this way, the design can be tested and adapted as necessary. The distance between two parallel laterals may vary between 50 and 150 m in permeable soils. In pervious clay soils, spacings of 20-50 m are common; in heavy clay soils and certain silt loams, spacings of 10-20 m are frequently required. In irrigated lands with an arid climate, the drain spacings are usually much wider than under rainier conditions owing to smaller discharges of the drains.

Drain Slope and Head Loss in the Pipes

The cost per unit length of installed field drains C_u is related closely to the drain diameter. This diameter depends on the expected outflow and on the available hydraulic head difference along the drain. Consequently, the drainpipe might be constructed without any slope. However, for practical reasons and cost-saving considerations, slopes are designed as high as possible in order to minimise the drain diameter. In sloping lands, drains can be laid parallel to the ground slope, especially where the surface has been graded. Thus, the pipe depth is maintained along the drain.

The usual criterion for sloping drains is that, at design discharge, no water is standing above the drain at its upstream end. However, interceptor drains, intended to collect and remove seepage water entering the top of the field, should follow the groundwater or soil surface contours. In flat lands, a shallower drain depth of the upper end of the drain must be chosen in order to maintain a minimum slope. However, very small slopes are possible, if the drains are constructed carefully and are sometimes used if subirrigation is to be practised.

In such horizontal drains, water must be allowed to temporarily stand above the drain in wet times, which by itself is not a problem as long as it remains deep enough below the soil surface. The argument that slope is needed to transport sediments out of the lateral is valid only at slopes of more than 1 percent, which are seldom possible in flat lands where drain slopes are usually in the range of 0.1 to 0.3 percent. Such a flat slope is not enough to remove incoming soil by the water flow.

Therefore, precautions against clogging are needed, i.e. careful construction of the drains and, in many cases, the use of protective drain envelopes. However, horizontal drains are not recommended because the installation tolerances are never negligible even where the drainage machine is equipped with a laser device. In practice, minimum slopes of 0.07 percent or in extreme cases 0.05 percent can be considered.

Drain Diameter

The design of the drain diameter should take into account the available diameters and the costs thereof. As cost increases with diameter, finances play a role in the choice of diameter. In designing the drain diameter, the total head loss in the drain during a very wet period H is considered. It is often required that, at design discharge, no water be standing above the upstream end of the drain. Therefore, with a slope of 0.2 percent and a length of 250 m, the available head for pipe flow is 0.50 m. If in flat land the drain outlet is 1.50 m below the surface, the depth of the drain at the upstream end will only be 1.00 m.

With an allowed head loss of 0.50 m, there will be no water above the pipe. In this case, drain slope and allowed head loss are the same. However, the same drain with the same outlet depth, but with a slope of 0.10 percent, has an upstream depth of 1.25 m below surface. With an allowed head loss of 0.50 m, there will be 0.25 m of water standing above the drain at design discharge, but the depth of this water will also be 1.00 m.

The same reasoning applies to any slope below 0.2 percent and even for a horizontal drain. This example shows that there is no direct relation between drain slope and allowed head loss. Therefore, the allowed head loss in the pipes will be taken as an input for calculations of the required drain diameters. This head loss determines the groundwater depth near the drain during critical times at the least favourable places.

The diameter of lateral and collector drains can be calculated using various formulae, which are based on the laws for pipe flow. These calculations are different for smooth and corrugated pipes, because of pipe roughness. The available head loss at design discharge and the amounts to be drained under that condition form the base for calculations concerning pipe diameters.

Drainage Materials

Corrugated Plastic Pipes

Corrugated plastic pipes with adequate perforations are most frequently used as field drains because of their flexibility, low weight and their suitability for mechanical installation, even for a drain depth of 2.5 m and more. Polyvinyl chloride (PVC) is commonly used in Europe, and polyethylene (PE) pipes are commonly used in North America, but both are technically suitable.

Although PE material is less resistant to soil loading than PVC and is sensitive to deformation at high temperatures, it is more resistant to ultraviolet radiation during storage and handling, and is less brittle at temperatures below 3 °C. However, the choice is usually based on availability and price considerations. Water enters into the drainpipe through perforations. These openings are uniformly distributed in at least four rows.

The perforated area varies from 1 to 3 percent of the total pipe surface area. Where the perforations are circular, diameters range from 0.6 to 2 mm. Elongated openings have a length of about 5 mm. In Europe, the perforation area should be at least 1 200 mm^2 per metre of pipe. Baked clay or concrete pipes about 30 cm long are still sometimes used, the former for field drains, and the latter mostly for large collector drains, especially where the required diameter is more than 200 mm. These pipes may be considered as "technically smooth".

Clay tiles have a circular cross-section with an inside diameter of 50-200 mm. For collectors, the inside diameters of concrete pipes range from 100 mm upwards. Where the diameter is more than 300 mm, reinforced concrete should be used. Where the sulphate content of the groundwater is high, it is necessary to use high-density cement resistant to gypsum. Drainage pipes should fulfil technical specifications that are verified in laboratories before installation. For plastic pipes, these specifications include impact resistance, weight, flexibility, coilability, opening characteristics and hydraulic characteristics.

Pipe Accessories and Protection Structures

At the upstream end of the drain, caps are used to prevent the entry of soil particles. Snap-on couplers are used to connect plastic pipes of the same diameter, and plastic reducers are used where the pipes are of different sizes. Where couplers and end caps are not available, the drainpipes can be manipulated in the field to fulfil the same functions. Rigid pipes, of sufficient length to prevent the penetration of roots of perennial plants growing on the ditch bank, are used as outlets. These pipes are also used where a drain crosses unstable soil, or a row of trees that may cause root intrusion.

Envelopes

To prevent soil intrusion in unstable silt and sandy soils, drainage pipes should be surrounded by envelope material. Envelope material can be made of: fine well-graded gravel; pre-wrapped organic materials, such as peat, or natural fibres, such as coconut fibres; or woven and non-woven synthetic materials, such as granular polystyrene and fibrous polypropylene. In soils consisting of stable clays at drain depth, such envelopes may often be omitted, which reduces drainage costs.

Envelopes prevent the entrance of soil particles, but they also promote the flow of water into the drain. A good envelope conveys water to the perforations, thus considerably reducing the entrance resistance. Moreover, voluminous envelopes increase the effective radius of the drain, from the pipe radius to that of the pipe plus envelope thickness. This further promotes water flow and improves the hydraulic efficiency of the drain. In addition to the entrance resistance restriction by soil clogging, drainage pipes have to face other problems, such as clogging of the pipe openings by penetration of roots into the pipe, by biochemical processes such as ochre formation, and by precipitation of less-soluble salts, such as gypsum and carbonates, which are difficult to prevent.

It is not easy to predict the need for an envelope but tentative prediction criteria are available. These criteria are based on clay content, soil particle size distribution, and salt and sodium content of the soil solution. Fine, well-graded gravel forms an excellent envelope, but the high cost of transport and installation constrains its use in practice. Organic fibres may decompose with time. Therefore, synthetic envelopes, such as pre-wrapped loose materials and geotextiles with appropriate opening sizes, are in widespread use.

Envelopes should also fulfil technical specifications, such as: thickness, mass per unit area, characteristic opening size and retention criteria, hydraulic conductivity and water repellence, and some mechanical properties. Guidelines for predicting the need for envelopes and for selection of the appropriate material are available, but the selected material must be field-tested for local conditions. Requirements for envelopes used for wrapped pipes are also included in the draft European standard.

Auxiliary Structures

Where singular subsurface drainage systems are used, a rigid outlet pipe is necessary. The rigid pipe should be long enough for water to flow directly into the outlet drain ditch water in order to prevent erosion of the ditch bank and to impede clogging by roots of bank vegetation. As these pipes hamper mechanical ditch cleaning, the bank may also be protected by concrete or plastic chutes.

In composite subsurface drainage systems, cross-connectors, T-pieces and elbows are used to join buried laterals and collectors. Junction boxes or fittings are used to connect pipes where the diameter or the slope of the pipe changes. Where inspection and cleaning are required, maintenance hatches replace junction boxes. Blind and surface inlets can be used to evacuate surface water through subsurface drainage systems. However, provision should be made to prevent entry of trash and eroded soil by using appropriate envelope material.

Interception Drains

Inflow from higher places or from leaky irrigation canals can sometimes be captured by interceptor drains, especially where it passes through relatively shallow aquifers. The effect of interception drainage is only significant if the impermeable layer is about at the drain depth. Otherwise, the effect is roughly only proportional to the percentage that the interceptor drain depth is of the thickness between the phreatic level and the impermeable layer. Interceptor drains can take the form of pipes or open ditches. However, with the latter, the stability of the side slopes is often problematic where large volumes are to be captured. Better solutions are gravel-filled trenches provided with a suitable pipe of sufficient capacity to carry the discharge.

Vertical Drainage

Vertical drainage is achieved by an array of properly spaced pumped wells that lower the head in an underlying aquifer and lower the water table. Vertical drainage can be used successfully under special physical circumstances:

- the presence of a good aquifer underneath (unconfined or semi-confined), so that wells give a good yield;
- a fair connection between the soil to be drained and the aquifer, so that the lowered head in the aquifer results in a lower groundwater table.

The layers between the aquifer and soil to be drained must be permeable enough to convey the recharge of the groundwater by rainfall and irrigation losses to the aquifer. In other words, the resistance between groundwater and aquifer must not be too high; ÿ the system should be sustainable. The aquifer should not be pumped dry. Where the water is to be used for irrigation or for municipal supply, a suitable quality is required that must not deteriorate rapidly with time. This sometimes occurs because vertical drainage may attract salt from deeper layers. As constant pumping is needed, the O&M costs are rather high.

This leads to the following economic restrictions:

- The method is only economically viable where the pumped water is fresh and can be used for the intended purposes. However, mixing with better quality waters can sometimes be a solution where undiluted use is not allowed.
- Where the water is too salty, it causes disposal problems in the project area that need special provisions. These

add to the costs, making vertical drainage still more uneconomic in these cases.

— The O&M costs and complexities of relatively dense well-fields limit the application of vertical drainage.

Despite these constraints, the method is applied widely in some areas where the soil and aquifer conditions are favourable and where the pumped water can be used. In such areas, it has often led to a depletion of the aquifer and sometimes to extraction of salts from deeper layers. Vertical drainage may also be an option in locations with severe seepage problems. Here, pumping is not always needed, because of overpressure in the aquifer. Where technically feasible, vertical extensions of a horizontal drainage system may be a cheap substitute. Relief wells consist of vertical wells that reach slightly into the aquifer.

In a drain trench, vertical boreholes are made into the aquifer and provided with blind-ended perforated pipes as well casings. They are usually made of corrugated plastic and are the same as the drain itself. These pipes are connected with the horizontal laterals by T-junctions. The method has been successful in several cases. However, the extra discharge of water may be a burden for the outlet system, and its salinity may harm downstream users.

References

Boussinesq, M.J., "Recherches théoriques sur l'écoulement des nappes d'eau infiltreés dansle sol et sur le débit des sources," *J. Math. Pures Appl.*, 10 : 1–78, 1904.

Boonstra, J., "Estimating peak runoff rates," *In* H.P. Ritzema, ed., *Drainage principles and applications*, 1994, pp. 111–143. 2nd edition, ILRI Publication 16, Wageningen, The Netherlands, ILRI.

Bruggeman, G.A., "Analytical solutions of geohydrological problems," *Dev. Wat. Sci.*, 46: 331, 1999.

Guyon, G., "Considérations sur le drainage des nappes, Théorie et Experimentation," *Bull.Tech. Gén. Rur.*, 1966, 79.

Moody, W.T., "Nonlinear differential equation of drain spacing," *J. Irrig. Drain. Div. ASCE*, 92(IR2): 1–9, 1967.

Soil Conservation Service (SCS)., "Hydrology," *National Engineering Handbook Section 4*, Washington, DC, USDA, 1972.

Smedema, L.K., Vlotman, W.F. & Rycroft, D.W., *Modern land drainage. Planning, design and management of agricultural drainage systems,* Leiden, The Netherlands, A.A. Balkema Publishers, Taylor&Francis, 2004, 446 pp.

Van Drecht, G., *Calculation of steady groundwater flow to ditches,* RID-Mededelingen 83-3, In Dutch with English abstract, The Netherlands, National Institute for Water Supply, 1983.

10

Installation and Maintenance of Drainage Materials

Installation Procedures

Manual installation of drains and installation with backhoe machines are a valid option for small drainage projects. Backhoes make wider trenches than drainage machines commonly used in large projects. They are also used for wide and deep excavations for large collectors. Drainage machines either make narrow trenches in which the drains are laid or they put the drain directly into the ground.

Trenching machines are either wheel or chain trenchers. They are appropriate for a wide range of working depths and widths. Trenchless machines can be classified in either vertical or V-ploughs. The trenchless installation method, however, has some practical limitations with respect to drain types, drain sizes, gravel application and installation depth. Therefore, trenchless drainage has not yet been widely implemented in irrigated areas.

Installing drains by manual labour or with classic excavators requires a series of successive operations: excavating the trench, installing the pipe, applying the envelope material and backfilling the trench. These operations are done simultaneously by trenching machines. Sometimes, backfilling is done by a separate auger or blade on a tractor. Backfilling can also be done by an implement, attached on the drainage machine when driving backward to begin excavating a new trench. Contemporary drainage machines are equipped with laser grade control, which has significantly contributed to the efficiency and accuracy in the installation of subsurface drains.

The maximum digging speed, however, should be adjusted to the speed of the hydraulic system that is used for automatic depth regulation, otherwise the installation accuracy will be poor. Although a certain deviation from the design grade can be tolerated, it should not exceed half the pipe diameter. Larger deviations promote air locks in high and sedimentation in low places, which obstruct water movement through the drain. Similarly, drain sections with a reverse grade cannot be tolerated.

Blinding

Since the risk of sedimentation is largest during installation and in the immediate subsequent period as long as the backfill has not settled and stabilised, drains are normally covered with friable topsoil to create a stable and highly permeable soil surround, and to preserve the alignment. Therefore trenching machines are equipped with cutters to bring a layer of topsoil or soil from another suitable layer from the sides of the trench on top of the drain. Its thickness should be at least 100 to 250 mm, depending on the drain diameter.

Granular envelope material can also be used to achieve a highly permeable drain surround and to prevent vertical and horizontal displacement once the pipe is installed. Any envelope material to be used must be in place around the pipe before blinding is done. Blinding, the initial covering of the drain with topsoil, is not recommended when organic envelopes are used, because topsoil with organic matter and intensive microbiological activity enhances the risk of microbiological decomposition of these envelopes. In such cases, soil from another suitable layer, with low organic matter, can be used for blinding. Further backfilling of the trench should be done as soon as possible and, at the latest, at the end of each day if there is a risk of surface water entering the trench.

Soil Cohesion

Since soil cohesion is strongly correlated with its water content, installation of the drainage system should preferably be done in unsaturated soil conditions with the water table below installation depth and outside periods of general wetness. In addition, the backfill should have settled before heavy rain or irrigation. In some situations, however, these conditions are not, or cannot be fulfilled. Drainage installation in wet conditions is discouraged, yet it is not always possible to drain under favourable or ideal circumstances. When cohesionless soils are drained in saturated conditions, an envelope must be wrapped immediately around the drain and the drain covered with backfill material before the liquid sand flows into the trench.

Caving of the trench wall, which often occurs in cohesionless or low cohesive soils, may damage and/or displace the drain. In every case, the drain and the envelope should be in place before the trench box has passed. Possibly, a longer trench shield may be used to protect a greater length of the trench. The drain should be blinded immediately. Simultaneous and instantaneous backfilling will help to prevent trench wall failure. However, the trench may collapse as soon as the trench box has passed and, therefore, a chute should be provided at the end of the trench box to convey the caving soil down to the top of the drain in order to avoid damage by falling clods and stones.

In cohesionless soils, drainage machines should be kept moving at all times. If not, fluid sand is likely to enter the trench box and cause problems with sedimentation as well as with alignment and grade of drains. Many problems, encountered with trenchers or backhoe excavators in saturated cohesionless soils, can be avoided by trenchless drainage installation. Drainage of physically stable, well-structured soils under general wetness may destroy the soil structure during excavation and create a less permeable trench backfill. Moreover, such conditions also promote mineral clogging of pipe and envelope.

In any case, the use of an envelope cannot compensate for the 'adversely affected' soil conditions. Every effort should be made to preserve the existing soil structure and to protect the drain from soil failure. Adjusting the forward speed of the machine can be done to limit the destruction of the soil structure. Observation of the condition of the excavated soil can be a guide to the proper machine speed.

The machine should move fast enough to preserve the structure of the soil and not turn the excavated soil into slurry. Structural deterioration of an originally stable, well-structured soil can be avoided with trenchless drainage installation. The functioning of drains installed with the trenchless technique depends very much on the changes in soil structure brought about by the passing of the blade. This depends on the soil, the circumstances and the depth.

Drainage of clay soils in wet conditions will unavoidably result in smearing and reduction of the hydraulic conductivity where the machine has physical contact with the soil. Drainage of cohesive soils in wet conditions must be avoided, regardless of the available drainage machine.

The installation conditions for laterals of a composite drainage system in saturated soil are improved if the time span between the installation of "permeable" collectors and installation of the laterals is long enough. This is because much of the local groundwater has the opportunity to drain out before the laterals are installed. In severe cases, where the construction of collectors is difficult because of quicksand, a temporary drain may be helpful. It is usually far cheaper than using well-points.

Backfilling and Finishing of Trenches

Backfilling and finishing of trenches should ensure a minimum of later land subsidence and preclude the occurrence of piping. The piping phenomenon may occur as a result of internal erosion of trench backfill by water flowing from the soil surface directly to the drains through the loose backfill material. This is crucial in irrigated lands, where irrigation water that can flow freely through the trench or drain plough fissures into the drainpipe, will dramatically lower the irrigation efficiency.

Furthermore, soil piping may cause soil material to be carried by the flowing water into the drain, creating sinkholes at the soil surface and/or mineral clogging of drains and envelopes, if present. Proper backfilling of the trench or plough fissures is therefore essential. It is easier to backfill and compact V-plough fissures than trenches. Fissures, created by vertical ploughs cause the most problems. Neither heavy loads, nor significant flooding should be imposed on newly installed drains until the soil in the trench is consolidated.

The loose backfill material will settle naturally with time. Since backfilling is usually done with a tractor equipped with a dozer blade, passage of the tractor wheel over the backfilled trench, filling it up, and running over it again will speed up the process, yet care must be taken to avoid crushing the pipe. This procedure ensures that only the top part of the trench backfill is compacted, and that the deeper part of the backfill retains a good permeability and a low entrance resistance. In case of trenchless drain installation with a vertical plough, compaction of the upper part of the disturbed soil is equally important.

A common procedure is that one track of the drainage machine runs over the drain line on its way back to the outlet drain to begin installing the next lateral. In dry soil, the rate of compaction following this procedure may not be sufficient. Application of irrigation water to unconsolidated material in trenches to settle the backfill is a practice that should be done very cautiously, however. If a field is to be flood irrigated before the trench backfill is consolidated, direct entry of uncontrolled surface water into the trench should be avoided by raising temporary ridges along both sides of the trench.

Guidelines with Respect to Drainpipes

Trenching machines can install clay, concrete, or plastic pipes. Clay and concrete pipes are manually placed on a chute that conveys the tiles down into the trench shield where they automatically move into the right position on the bottom of the trench. The tiles should be installed in the trench in such a way that a perfect junction between drains is obtained. For drains of larger sizes, an inspector, standing or sitting in the shield, checks for correct laying.

The maximum gap between drains may not be more than 3 mm except for sandy soils or soils with a sandy layer on drain depth where it should be not more than 2d85. Clay and concrete tiles without gravel or appropriate synthetic envelopes are not recommended in cohesionless fine sand. Plastic drains are normally fed through a. conducting pipe, mounted just behind (wheel trencher) or above (chain trencher) the digging mechanism of the trencher. Trenchless machines have been developed to install only corrugated drains of not too large a diameter. They should not be installed with a curvature radius less than five times the pipe diameter, particularly if the pipe is wrapped with an envelope.

For machine installation, the quality of drainpipes is of utmost importance. Drainpipes with fissures, cracks or other visible shortcomings and badly formed pipes or torn envelope material, which do not allow a proper installation or assure a reliable performance, should not be used. Furthermore, all drains and collectors must be closed at the upward end to avoid soil invasion. Failures that may occur during installation of corrugated drains are crushed or collapsed pipes, twisted pipe sections, couplings pulled apart and snappedoff pipes. In such cases, the discharge is obstructed. Although the water may finally find its way through the soil to a properly functioning downstream part of the drain and to neighbouring drains, stagnation occurs.

Upstream the blockage, water may stand above the drain and a higher groundwater table will result. Coils of smaller diameter pipes are usually carried on a reel on either trenching or trenchless machine and wound off as installation proceeds. Larger diameter pipes are usually laid out on the field beforehand, and then guided through the trenching machine. Excessive pulling can result in connections becoming loose or pipes breaking off. During the uncoiling of the pipe, pipe breakage can be easily overlooked, yet the missing piece of drain will cause local wetness. Therefore, trenchless drainage machines must be equipped with guides to facilitate smooth entrance of the drainpipe into the feeder tube.

Gravel envelope application can entail substantial, undesirable elongation of the drainpipe if the gravel does not flow smoothly downward through the supply tube. While cleaning corrugated PVC drains by jetting, it is sometimes observed that drains were not laid in a straight line, but spiralled slightly. This phenomenon is attributed to the tension in the pipe material generated in the unwinding of the rolls at installation, and may enhance the development of unwanted airlocks inside the drain.

PVC pipes should not be installed at temperatures below 3°C because of their brittleness at low temperatures. Storage at temperatures exceeding 40°C for PE and 80°C for PVC pipes, as well as installation at temperatures above 40°C should be avoided in order to prevent pipe deformation as a result of load and longitudinal stress. Exposure to UV rays of solar radiation also affects the strength properties of corrugated plastic pipes. Stored pipes should therefore be protected from the influence of direct sunlight if not installed within one week or one month after delivery.

Envelope Material

Whatever envelope material is used, and by whatever method it is installed, envelopes must fully surround a drainpipe, unless the drain is installed on an impervious layer. An envelope merely on top of a drain does not suffice because mineral clogging also occurs from underneath if water enters the drain from all around. Bulky envelopes can be spread out by hand in the bottom of the trench before the pipe is placed, but this is only possible in stable soil where trench walls do not collapse. If drains are laid by hand and a layer of the bulky envelope should surround the drain, the envelope is placed on the bottom of the trench and levelled first. Next, the drain is installed and covered further with bulky envelope to the required height. This also holds for machine installation of drains with a bulky envelope.

Envelope strips, delivered on rolls, should be applied below and on top of the drain. The material at the bottom needs not necessarily be the same as the material on the top. Prewrapped drains, however, are preferred since they protect drains from all sides, and offer a greater safety than bulky envelopes or envelope strips can do. Envelopes that are good and reliable, however, will only be successful if properly installed under favourable physical soil and weather conditions. Slurry in the bottom of a trench will cause immediate and complete failure of the envelope material and hence of the drain.

The general use of gravel envelopes has decreased continuously in spite of all efforts to mechanise and perfect installation by e.g. introducing a gravel auger at the end of the trench box. This gravel auger reduces pipe stretch but gravel-feeding problems are still not completely solved. Theoretically, it is also possible to apply gravel with the vertical drain plough as well as with the V-plough. However, the risk of stagnation of gravel in the supply tube of the machines makes the trenchless technique less suitable for gravel installation. The installation of gravel remains a difficult and labour-intensive operation. Practical experience shows shortcomings causing base soil intrusion and pipe siltation. The major shortcomings are:

- — segregation during transportation and installation;
- — flow problems in the supply tube;
- — unequal distribution around the drainpipe; and
- — accidental incorporation of soil into the gravel on the bottom of the stockpile.

Coarse, well-graded sand can also be used as a drain envelope. However, the shear resistance of sand, especially if it is not completely dry, will hamper mechanical installation even more seriously than gravel does. Organic and synthetic envelopes, prewrapped around corrugated drainpipes can be installed adequately with both trenching and trenchless machines. They are however prone to damage, caused by transport and/or rapid machine installation, especially when materials of inferior quality are used or when the pipe is not carefully wrapped.

In order to avoid local spots of soil particle invasion, prewrapped envelopes cover the entire drain circumference. Furthermore, they should not be damaged during handling and installation. Therefore, the layer of loose material before wrapping should be sufficiently thick and as uniform as possible to avoid open spots. Geotextiles that are used for the wrapping of drainpipes are usually supplied on rolls.

The sheets should be wide enough to facilitate adequate overlap so that the pipes are completely wrapped, without open joints. If both longitudinal edges of a geotextile sheet are sewn, the sheet should be wide enough to facilitate this. If a geotextile sock is pulled manually over the drain laid out on the field, both the geotextile and the seam, if any, should be strong enough to resist this handling without damage.

Geotextiles usually have adequate mechanical strength to resist mechanical loads during installation. Machine installation requires adequate drainage materials to assure a straightforward installation and a proper drainage performance. Therefore high-quality materials are required and their properties need be checked prior to installation according to well-considered standard specifications. Quality standards of drainpipes and drain envelopes are therefore of paramount importance. Neither PLMs nor prewrapped geotextiles show particular problems during installation with both trenching and trenchless machines. Their light weight makes them suitable in soft soils where the use of gravel creates problems because of the weight of the gravel.

Maintenance of Drain Pipes

Jet Flushing Technique

Maintenance is obvious when there is severe clogging. If done regularly it may extend the service life of the system and enhance its performance. In case of light obstructions in pipes dry rodding may be helpful: a long series of coupled rods, with a scratcher at the end, is pushed into the drain and removed later. If done during a period of considerable discharge, the loosened materials will be discharged. For more serious forms of clogging, jet flushing has to be used. Jet flushing is a technique used to remove clogging and precipitating agents from drainpipes through the impact of water jets. More particularly, the functions of jet flushing are:

- — lifting of blockages inside the pipe drain;
- — removal of deposits from the inner wall surface of the drain;
- — cleaning of clogged perforations;
- — removal of loose smaller roots of agricultural crops and weeds; and
- — supply of sufficient water to carry the loosened agents, including sand and clay particles towards the drain outlet.

Ideally, the water that discharges from the drain evacuates the major part of the clogging agents. Particles, larger than approximately 75 μm may be dislodged, yet are generally too heavy to be removed from the drain. It is not clear to what extent pipe perforations can be cleaned efficiently and non-destructively. It is assumed that jet flushing has a negligible effect on clogged envelopes. A typical jetting device is operated from the power takeoff of an agricultural tractor. It consists of a pump, a suction pump inlet, and a reel with a 200-400 m long pressure hose fitted with a nozzle.

The nozzle is fed into the pipe drain from the downstream end. Therefore, the pressure hose is pointed to the drain outlet with the help of an adjustable hose guide. Access of the outlets of laterals is easy if they discharge into open collector ditches. Contrary to these singular drainage systems, as common in humid temperate zones, drainage systems in semiarid countries often have a composite layout, whereby laterals discharge into pipe collectors instead of open collectors. If the junctions between laterals and collectors are located at manholes, these can be used to accept a jetting hose, provided that the diameter of the manhole is at least 0.3 m. In some countries, e.g. Egypt, laterals are accessible at their upstream end.

On average, jetting requires 1-2 m^3 of water per 100 m of drain. The water can be pumped from a drainage ditch, an irrigation supply canal, or a tanker must supply it. Saline water is a harsh and corrosive environment for flushing machines. If saline water must be used, the flushing machine should be made of high quality salt resistant machine parts.

The use of salt water for flushing must be avoided: it damages the soil structure around the drain and it is harmful for the machine. During the jetting procedure, the nozzle must be inserted into the pipe as fast as possible. The pulsating action of the piston pump enhances the forward movement of the nozzle. After the nozzle has reached the upstream end of the drain, the hose is retreated by reeling, at a steady pace of approximately 0.3 m/s while pumping continues.

The cleaning action is influenced by the cleaning force, the angle of attack of the water jets, the duration of cleaning, the water temperature and the use of chemicals. The cleaning force is proportional to the flow rate times the square root of the water pressure at the nozzle. Environmental restrictions as well as cost considerations generally preclude the use of chemicals while jetting. A balance must be found between the pressure and the flow velocity of the water jets coming from the nozzle, preferably on site. The optimum ratio is likely to depend on the inside diameter of the drains; however, no data are available to support this assumption.

On many commercial jet flushing units, the ratio between flow rate and pressure can be adjusted. Flow rates are adjusted by changing the pumping speed. The water pressure is adjusted by selecting an appropriate nozzle. Jet flushing will temporarily increase the water pressure in the drainpipe and thus in the surrounding soil, possibly affecting soil stability around the drain. The increased water pressure causes a reduction of cohesive forces between soil particles, which may lead to instant and hazardous quicksand conditions.

Notably in weakly cohesive soils, there is a risk of the development of quicksand. After the nozzle has passed, structureless soil material may flow into the pipe. In addition, the hydraulic conductivity of the soil may be adversely affected. Regardless of the discharge from the nozzle, dislodged substances are more easily evacuated from small than large diameter drains due to the higher flow velocities in the smaller diameter pipes. As far as the water pressure is concerned, three categories of jet flushing units are being manufactured:

— high pressure equipment : > 100 bar at the pump;

— medium pressure equipment : 20-35 bar at the pump;

— low pressure equipment : < 20 bar at the pump.

High-pressure units cannot be recommended, because empirical experience evidenced that this type of flushing machine destabilises the soil around the drain and destroys its structure. Water pressure at the nozzle is approximately 50 percent of the pressure at the pump. Hydraulic data of nozzle, pump pressure, and flow rates provided by a commercial flushing unit manufacturer for a flexible hose with an inside diameter of 20 mm and a length of 300 m.

The highlighted line contains recommended figures. The maximum flow of water that can be employed depends on the cross section of the drain. Empirically it was found that a discharge of approximately 70 l/min is satisfactory for 50 to 70 mm pipe diameters. Such discharges are indeed realised with the highly popular medium pressure units.

Higher discharges may force too much water through the pipe perforations, which is hazardous for the envelope and the structure of the abutting soil. The cost/benefit effects of regular maintenance of drains by jet flushing are hard to quantify. Still, some figures may be informative. The cost of jet flushing in The Netherlands, at medium pressure, is approximately US $0.15 per m of drain which is 12 percent of the installation cost of $1.25 per m. With a typical drain length of 800 m per hectare and a flushing frequency of once in every three years, the annual cost amounts to $40 per hectare per year. The average annual gross yield of arable land is approximately $2500 per hectare. The calculated maintenance cost is therefore less than 2 percent of the annual gross yield.

Dry Rodding and Jetting of Drains

Dry rodding and jetting of drains are useful for removing ochreous substances but generally not for removing roots from drains, with the exception of loose, tiny ones. Before jetting, some drains should be examined internally first, e.g. with a miniature video camera, in order to check the kind of clogging and to assess the jetting efficiency. In case of ochreous substances, preventive jetting may be useful in order to prevent total blocking of pipe perforations.

Ochre is a soft substance when precipitating, but becomes dense and sticky with time, making it difficult to remove. Jetting cannot generally reopen pipe perforations that were clogged with encrusted ochreous substances. Ochre deposits should therefore be removed before drying out by frequent flushing with medium pressure. Based on recently acquired experience in The Netherlands, this recommendation is nowadays relaxed somewhat in the sense that flushing is recommended only if the ochre deposits do noticeably impede proper functioning of the drain. This recommendation also holds for other kinds of microbiological deposits inside drains.

The following conditions may enhance the risk of drain sedimentation through jetting:

- the use of high pressure equipment;
- jetting shortly after drain installation (soil not yet settled nor stabilised);

— damaged pipes and/or decomposed envelopes;
— non-cohesive and weakly-cohesive soils; and
— slow pace of movement or (temporary) blockage of the nozzle.

In The Netherlands, approximately 600000 hectares of agricultural lands are provided with a subsurface drainage system. No precise data about the area periodically flushed is available. In 1998, the number of flushing units in operation was estimated at several thousands, so a considerable area is regularly maintained. The medium pressure unit is by far the most widely used. In the past, jet flushing has been reported to have a positive effect on drain performance in a pilot area, where drains were prone to excessive biochemical clogging due to intense upward seepage of ferrous groundwater.

As long as the drains wère jetted periodically, the drainage system met the design criteria in terms of drawdown of groundwater and discharge. After jetting was discontinued, the plots suffered from waterlogging. Van Hoorn and Bouma investigated the effect of jetting on drains, installed in clay soils, which had been submerged regularly and clogged by mineral particles and biochemical substances. The effect was quite positive. At another pilot area in The Netherlands with comparable conditions, however, Huinink established that drain performance could not be restored, despite the implementation of an extensive jetting project.

Experiences with high-pressure equipment in northwestern Europe are unfavourable, while substantial pipe sedimentation is occasionally reported with intermediate pressure equipment. Practical experience of farmers and contractors learned that flushing with high pressures enhances sedimentation rates. The next flushing had to be done sooner than in case medium or low pressure was used. Around 1980, therefore, the use of high-pressure equipment was gradually discontinued. During the nineties, the frequency of jet flushing as advised to the farmer varied from annually to once in every five years.

During this decade, farmers have gradually become somewhat suspicious towards jetting of drains. Intense monitoring of drain performance in various pilot areas revealed that the assumed beneficial effects were not so obvious as was assumed for a long time. If any improvement in drain performance could be noticed at all, it would generally last for a very short time.

This fact has induced some reluctance towards preventive jetting of drains. Drainage experts nowadays give the following advice to the farmers: do not jet any drain as a form of preventive maintenance, unless there is a substantial risk of ochre clogging. On the other hand, jetting is useful if the performance of drains has significantly deteriorated, as observed by the farmer.

Drains, prewrapped with suitable and lasting envelopes should however be practically maintenance free. A likewise observation was made in the United States some 20 years earlier. Because of this development, the number of Dutch manufacturers of high and medium pressure equipment went down from six in 1991 to two in 1998. Comparatively simple low pressure jetting equipment is however manufactured at various locations.

References

Ami, S. R., *Drainage Pipe Testing Manual,* Canadian International Development Agency (CIDA), Hull, Quebec, Canada, 1987.

Antheunisse, J., "Biological and chemical inhibition of ligno-cellulose decay by white rot basidiomycetes," *J. Gen. Appl. Microbiol.,* 29: 257-269, 1983.

Bertram, G.E., "An experimental investigation of protective filters," *Harvard Soil Mechanics Series 7,* Pierce Hall Harvard University, 1940.

Chapman, D., "Ditches, dykes and deep-drainage," *Young Farmer's Club Booklet 29,* Evans Brothers LTD, London, 1956.

Childs, E.C. and Youngs, E.G., "The nature of the drain channel as a factor in the design of a landdrainage system," *Journ. Soil Sci.* 9, 2: 316-331, 1958.

Dierickx, W., "Changes in specific properties of PVC drains under various storage conditions," *Agric. Water Manage,* 37, 2: 109-119, 1998a.

Irwin, R.W.; "Corrugated pipe flow rate from pipe geometry," *Journ. Irr. and Drainage Eng.*, 2: 237-241, 1984.

Lennoz-Gratin, C., "The use of geotextiles as drain envelopes in France in connection with mineral CLOGGING RISKS," *Geotextiles and Geomembranes* 5, 2: 71-89, 1987.

Nieuwenhuis, G.J.A. and Wesseling, J., "Effect of perforation and filter material on entrance resistance AND EFFECTIVE DIAMETER OF PLASTIC DRAIN PIPES," *Agric. Water Manage*, 2: 1-9, 1979.

SHERARD, J.L., DUNNIGAN, L.P. AND TALBOT, J.R., "BASIC PROPERTIES OF SAND AND GRAVEL FILTERS," *Journ. Geotech. Engr. Amer. Soc. Civil Eng.* 110, 6: 684-700, 1984A.

WESSELING, J. AND HOMMA, F., "ENTRANCE RESISTANCE OF PLASTIC DRAIN TUBES," *Neth. J. Agric. Sci.* 15: 170-182, 1967.

11

Evaluation of Irrigation Project

The diagnosis or appraisal of project performance provides the fundamental basis for designing modernisation strategies and plans. Thus, if it is not done properly, the whole modernisation process will probably be flawed and fail to yield the intended results. Appraisal of irrigation system performance should help in the identification of short-term, medium-term and long-term actions needed to improve its performance.

An appraisal or evaluation must be:

— systematic: conducted using clear, step-by-step procedures, well planned, and precise;

— objective: if done by different professionals, the results should not differ;

— timely and cost-effective (not taking too much time, and not too expensive);

— based on a minimum of data required for a thorough evaluation.

It should cover:

— all aspects that could influence actual water delivery service, including the physical infrastructure, water management practices, roles and responsibilities governing WUAs, budgets, and maintenance;

— all levels of the system.

A proper diagnosis or appraisal process should be based on a combination of:

— field inspections, for evaluating physical system and operations;
— interviews with the operators, managers and users, for evaluating management aspects;
— data analysis, for evaluating a water balance, service indicators and physical characteristics.

Conventionally, appraisals of irrigation systems often look at the big or overall picture and consider the inputs (water, labour, overall cost, etc.) and outputs (yield, cost recovery, etc.) of a system.

While the overall picture is important, it does not provide any insight into what parts or components of a system should be improved or changed in order to improve the service in a cost-effective manner. Therefore, a sound diagnosis should provide insights into the internal processes as well as outputs.

Internal and External Indicators

The internal indicators assess quantitatively the internal processes of an irrigation project. Internal indicators are related to operational procedures, the management and institutional setup, hardware of the system, water delivery service, etc (Table 1).

Table 1. Examples of internal and external indicators

Internal indicator	External indicator
Flow rate capacities	Command area efficiency
Reliability	Field irrigation efficiency
Flexibility	Production per unit of land (US$/ha)
Equity	Production per unit of water (US$/m3)

These indicators are necessary in order to have a comprehensive understanding of the processes that influence water delivery service and the overall performance of a system. Thus, they provide insight into what could or should be done in order to improve water delivery service and overall performance. The external indicators compare the inputs and outputs of an irrigation system in order to describe overall performance.

These indicators are expressions of various forms of efficiency, e.g. water-use efficiency, crop yield, and budget. They do not provide any detail on what internal processes lead to these outputs and what should be done in order to improve performance. However, they could be used for comparing the performance of different irrigation projects both nationally and internationally.

Once these external indicators have been computed, they can be used as a benchmark for monitoring the impacts of modernisation on improvements in overall performance.

Evaluating Irrigation Projects

An irrigation project can be appraised in many different ways incorporating all or some of the elements described above. The methodologies commonly used by researchers and evaluators of the system make use of checklists, detailed data collection and analysis, participatory rural appraisal (PRA) techniques, and detailed surveys. However, the use of these tools depends on the perspective with which diagnostic analysis is performed.

Traditionally, diagnostic procedures have focused on only one or two of components, e.g. equity in water delivery or institutional reforms, and only covered part of the system, e.g. one lateral. These limited-purpose diagnostic studies have usually been based on the collection of substantial field data and, thus, are time-consuming and expensive.

Field data collection is feasible for long-term research projects. However, for project appraisals and diagnosis for modernisation improvements, it is often necessary to evaluate the situation rapidly with whatever data are available.

FAO Approach to Irrigation System Appraisal

Rapid and focused examination of irrigation projects can give a reasonably accurate and pragmatic description of the current status of an irrigation system, and of the processes and hardware/infrastructure that in turn result in the present condition. It is on this basis, that FAO, together with the ITRC and the World Bank, developed a methodology/tool called the RAP with well-defined procedures for the rapid assessment of the performance of irrigation schemes.

The RAP allows for the identification of major actions that can be taken quickly in order to improve water delivery service (especially where the diagnosis is made in cooperation with the local irrigation authorities). It also helps in identifying long-term actions and the steps to be implemented in a modernisation plan.

Although irrigation systems can be evaluated and appraised using any or combinations of the above-mentioned methods, FAO recommends using the RAP because of its rapid nature, systematic procedures, and comprehensive approach, as it covers all the different components (physical, management and institutional) of an irrigation system.

Rapid Appraisal Procedure

The RAP was developed originally by the ITRC in the mid-1990s for a research programme financed by the World Bank on the evaluation of the impact on performance of the introduction of modern control and management practices in irrigation. Since its introduction, the RAP has been used successfully by FAO, the World Bank and other irrigation professionals for appraising projects in Asia, Latin America, and North Africa.

The conceptual framework of the RAP for the analysis of the performance of irrigation systems is based on the understanding that irrigation systems operate under a set of physical and institutional constraints and with a certain resource base.

Systems are analysed as a series of management levels, each level providing water delivery service through the internal management and control processes of the system to the next lower level, from the bulk water supply to the main canals down to the individual farm or field.

The service quality delivered at the interface between the management levels can be appraised in terms of its components (equity, flexibility and reliability) and accuracy of control and measurement, and it depends on a number of factors related to hardware design and management.

With a certain level of service provided to the farm, and under economic and agronomic constraints, farm management can achieve certain results. Symptoms of poor system performance and institutional constraints are manifested as social chaos, poor maintenance of infrastructure, inadequate cost recovery and weak WUAs.

The basic aims of the RAP are:

— assess the current performance and provide key indicators;
— analyse the O&M procedures;
— identify the bottlenecks and constraints in the system;
— identify options for improvements in performance.

The RAP can generally be completed within two weeks or less of fieldwork and desk work if some data are made available in advance by the system managers. A set of Excel spreadsheets in a workbook is developed in order to conduct the RAP.

These spreadsheets provide the evaluators with a range of questions related to the physical, management and water systems of an irrigation project that the evaluator has to answer. Based on the data and information input, a set of internal and external indicators is computed automatically.

The RAP has also been used as a foundation for benchmarking. The International Programme for Technology and Research in Irrigation and Drainage (IPTRID) defines benchmarking as a systematic process for achieving continued improvement in the irrigation sector through comparisons with relevant and achievable internal or external goals, norms and standards.

The overall aim of benchmarking is to improve the performance within an irrigation scheme by measuring it against desired targets and own mission and objectives. The benchmarking process should be a continuous series of measurement, analysis and changes to improve the performance of the schemes. Thus, the RAP becomes a tool for regular M&E of an irrigation project.

Physical Infrastructure of Irrigation System

The physical infrastructure or hardware (reservoirs, canals, diversion and distribution structures, etc.) of an irrigation system is the major physical asset of an irrigation authority or water service provider.

Keeping the infrastructure/hardware in reasonable shape and operating it properly is the only way to achieve water delivery targets, provided that the delivery targets are set realistically. The main items to examine while appraising the physical characteristics of a system are:

- assets: conveyance, diversion, control and other structures per kilometre;
- capacities: canals and other structures;
- maintenance levels;
- ease of operation of control structures;
- accuracy of water measurement structures;
- drainage infrastructure;
- communications infrastructure.

Project Management

The management arrangements, procedures, incentives, etc. of any irrigation system play a vital role in how it is operated. The ways in which decisions are made, communicated and implemented influence not only the way the system is managed but also the perceptions of users about how the performance of the system meets their needs. Often, operations, and thus water delivery service, could be improved significantly without much monetary investment by improving operational procedures, including for example the way control structures are manipulated. However, this often requires capacity development and appropriate targeted training of the office personnel and operators.

In order to identify improvements in the management of a project, it is necessary to appraise the following items:

— operation:
  — water allocation and distribution rules,
  — rules and procedures for operation,
  — stated vs actual policies and procedures,
  — the way structures are manipulated and operated - how changes are managed,
  — communication,
  — skills and resources of the staff at all levels;
— budget:
  — how realistic the budget is for the system operation to achieve set targets,
  — cost recovery - whether the system is able to pay for itself and invest in improvements as needed;
— institutional:
  — user satisfaction,
  — user involvement in decision-making - WUA.

Water Management

Water Delivery Service

Irrigation systems are composed of hydraulic layers, where each layer or level provides service to the next, lower level. Therefore, it is necessary to evaluate water delivery service at all levels. At each level in general and for water users in particular, it is very important to receive the required volume of water at the right time, thus adequacy, reliability and timeliness are crucial. However equity of water deliveries is also a critical target for managers. Therefore, adequacy.

reliability and equity indicators are often used for assessing water delivery service.

Other important indicators, particularly for modernisation, are flexibility (frequency, rate and duration) and measurement of volumes. Farmers can strategies and plan their cultivation and irrigation activities better where they can choose or at least predict the frequency, rate and duration of water delivery. Thus, the RAP computes the following indicators for assessing water delivery service at each level of an irrigation system:

— reliability,

— equity,

— flexibility,

— measurement of volumes.

Irrigation systems are often under increasing pressure to provide water for uses other than irrigation. In such cases, it is also necessary to evaluate the level of service required for these other uses.

Water Balance

A water balance provides an accounting of all the inflows and outflows within a defined boundary, as well as information about different water efficiencies. Thus, it provides a good assessment of existing constraints and opportunities for improvement. It helps set the stage for determining the level of water delivery service to be achieved and for designing appropriate allocation strategies.

The RAP includes a water balance at the system/project level for the rapid assessment of the external indicators and identification of the potential for water conservation. However, for regular monitoring and water management decisionmaking, a more detailed water balance is required.

Capacity Development for Diagnosis and Evaluation

Managers, engineers and national experts are not usually equipped to systematically evaluate the performance of irrigation projects and appraise modernisation improvements. Therefore, international experts are brought in at the initial phase for project appraisal. However, there is the risk that, once the project has been implemented and the international experts have gone, everyone can go back to "business as usual" and the project can return to its routine cycle of operation without any M&E of service.

Moreover, changing the mindsets of irrigation authorities from supply-oriented management to SOM requires substantial investment in the capacity building of managers, engineers, national experts and water users.

Even the well-documented procedures of the RAP require the adequate training of an experienced water resources professional. Experience has shown that successful application of the RAP requires:

- prior training and field experience in irrigation and drainage;
- specific training in the RAP techniques;
- follow-up support by trained experts when the evaluators begin their fieldwork.

Without investing in capacity building, modernisation projects will not yield the desired results. There is a need to raise the capacity of irrigation personnel in order to enable them to evaluate critically their own system and be able to appraise conditions objectively, and to propose and undertake improvements in consultation with the users.

Thus, it is critical to have capacity development programmes at project and national level with a view to promoting the adoption of effective irrigation modernisation strategies in support of agricultural development, increases in water productivity and IWRM.

Any modernisation programme undertaken without adequate associated capacity development programmes may fail to produce real improvements and may result in considerable amounts of money being wasted.

- — rotation schedules that are not enforced rigorously;
- — institutionally weak WUAs with responsibility for O&M of substantial portions of the project, but which have only minimal budgets;
- — severe inequity (tail-ender problems);
- — low collection rates for an irrigation service fee that is set well below actual costs;
- — phased implementation rehabilitation efforts, which have resulted in a mixture of different water control strategies and hardware (fully gated vs proportional flow division).

An RAP diagnostic evaluation was performed in different parts of the SMIS in two and a half days of intensive fieldwork. The results of the RAP quantified the performance of the SMIS in terms of the quality of water delivery service at each canal level in the system.

Internal indicators showed that only marginal improvements have been made in the most recent command area development. However, they demonstrated clearly that the design concept of proportional flow division does not provide the operational flexibility required for meeting demand variations (owing to rainfall, crop diversification, etc.).

In addition, a major deficiency of this design is the inequity that results from less than the full design capacity being achieved as a consequence of either low-flow conditions in the main canal or changes in the hydraulic characteristics of various canals caused by siltation, weed growth, etc. Although the new system has been in operation for one year, operators have already reacted by installing steel gates at proportional structures in order to regulate the flow in some tertiary canals.

The phased implementation of construction activities and institutional development in different stages of the SMIS has resulted in relatively better service in some parts of the project. However, it has also resulted indirectly in not enough attention being paid to overall issues such as how water is controlled in the main canal. One lesson of the SMIS RAP is that it is critical to ensure that the technical/ engineering details are correct before expecting any success in participatory management schemes.

The present operation of the CMC results in severe inequities in the "undeveloped" areas of the project. The design of the main canal cross regulators (manually operated, vertical steel gates with no side weirs) makes it difficult to maintain constant upstream water levels, which is compounded by the operation of the secondary canal offtakes. Water delivery service is relatively poor at all levels of the SMIS but worsens at the tertiary canal level, which is the interface where water users groups (WUGs) are supposed to take over O&M from the staff of the Department of Irrigation (DOI).

Part of the reason for the inadequate quality of service is related to the hydraulic characteristics of the cross regulators (manual undershot gates) in secondary and subsecondary canals.

References

Busman, J.D., *Optimizing control of surface irrigation using concurrent evaluation of infiltration*, PhD Dissertation, Agricultural and Irrigation Engineering, Utah State University, Logan, Utah. Unpublished document. 209p.

Doorenbos J. and W.O. Pruitt, *Crop water requirements*, Irrigation and Drainage Paper 24, Rome. 144p, 1977.

Garton, J.E., *Designing an automatic cut-back furrow irrigation system*, Oklahoma Agricultural Experiment Station, Bulletin B-651, Oklahoma State University, Stillwater, Oklahoma, 1966.

Gharbi, A., *Effect of flow fluctuations on free-draining and sloping furrow and border irrigation systems*, MS Thesis, Agricultural and Irrigation Engineering. Utah State University, Logan, Utah., Unpublished document, 123p, 1984.

Haise, H.R., Kruse, E.G., Payne, M.L,. and Duke, H.R., *Automation of surface irrigation: 15 years of USDA research and development at Fort Collins, Colorado*, USDA Production Research Report No. 179. US Government Printing Office, Washington DC, 1980.

Humpherys, A.S., *Surge flow surface irrigation: Section 3, Equipment,* Final report of Western Regional Project W-163, Utah Agricultural Experiment Station, Utah State University, Logan, Utah. 106p, 1987.

Ismail, S.M. and Westesen, G.L., *Surge flow border irrigation using an automatic gate,* Paper 84-2069 presented at the Winter Meeting of ASAE, Chicago, Illinois, 1984.

Jensen, M.E. (ed.), *Consumptive Use of Water and Irrigation Water Requirements,* American Society of Civil Engineers, New York. 215p, 1973.

Kemper, W.D., *et al., Cablegation system for irrigation: description, design, installation, and performance,* USDA-ARS Pub. 21, US Government Printing Office, Washington DC, 1985.

Bibliography

Busman, J.D., *Optimizing control of surface irrigation using concurrent evaluation of infiltration*, PhD Dissertation, Agricultural and Irrigation Engineering, Utah State University, Logan, Utah. Unpublished document. 209p.

D'Urso, G., Querner, E.P. and Morabito, J.A. "Integration of hydrological simulation models with remotely sensed data: an application to irrigation management", In: *Advances in Planning, Design and Management of Irrigation Systems as Related to Sustainable Land Use*, Vol. 2. J. Feyen *et al.* (eds.). Leuven, Belgium. pp. 463-472, 1992.

Dilip Kumar Majumdar, *Irrigation Water Management (Principles & Practices)*, Prentice Hall of India (P), Ltd.

Doorenbos J. and W.O. Pruitt, *Crop water requirements*, Irrigation and Drainage Paper 24, Rome. 144p, 1977.

Garton, J.E., *Designing an automatic cut-back furrow irrigation system*, Oklahoma Agricultural Experiment Station, Bulletin B-651, Oklahoma State University, Stillwater, Oklahoma, 1966.

Gharbi, A., *Effect of flow fluctuations on free-draining and sloping furrow and border irrigation systems*, MS Thesis, Agricultural and Irrigation Engineering. Utah State University, Logan, Utah., Unpublished document, 123p, 1984.

Irwin, R.W., "Corrugated pipe flow rate from pipe geometry," *J. Irr. Dr. Eng.*, 1984, 237–241.

Irwin, R.W., "Corrugated pipe flow rate from pipe geometry," *Journ. Irr. and Drainage Eng.*, 2: 237-241, 1984.

Lennoz-Gratin, C., "The use of geotextiles as drain envelopes in France in connection with mineral clogging risks," *Geotextiles and Geomembranes* 5, 2: 71-89, 1987.

Menenti, M., Visser, T.N.M., Morabito, J.A. and Drovandi, A., "Appraisal of irrigation performance with satellite data and geo-referenced information", In: *Irrigation Theory and Practice*. J.R. Rydzewsky and K. Ward (eds.). London. pp. 785-801, 1989.

Merriam, J.L. and Keller, J., *Farm irrigation system evaluation: A guide for management*, Department of Agricultural and Irrigation Engineering, Utah State University, Logan, Utah., 1978.

Nieuwenhuis, G.J.A. and Wesseling, J., "Effect of perforation and filter material on entrance resistance and effective diameter of plastic drain pipes," *Agric. Water Manage*, 2: 1-9, 1979.

Sherard, J.L., Dunnigan, L.P. and Talbot, J.R., "Basic Properties of Sand and Gravel Filters," *Journ. Geotech. Engr. Amer. Soc. Civil Eng.* 110, 6: 684-700, 1984a.

Van der Beken, A, De Troch, F., De Somer, M. & Zuidema, F.C., "The lateral inflow into submerged drains," *Bull. Int. Ass. Hydrol. Sci.*, 17: 273–289, 1972.

Wesseling, J. and Homma, F., "Entrance resistance of plastic drain tubes," *Neth. J. Agric. Sci.* 15: 170-182, 1967.

Zuidema, F.C. & Scholten J., "De afvoercapaciteit van geribbelde plastieken draineerbuizen," *Van Zee tot Land*, 49: 73–102, 1972.

Index

American Society for Testing and Materials (ASTM) 166

Basin irrigation method 57
Border irrigation method 57

Canal irrigation systems 84
Central ground water board 2
Central water commission 2
Chatra Main Canal (CMC) 264
Command Area Development Programme (CADP) 9
Cost-effective management 74
Consumptive Irrigation Requirement (CIR) 45
Conventional optimization techniques 113
Culturable Command Area (CCA) 5, 34
Curve Number (CN) 216

Department of Irrigation (DOI) 267
Drip irrigation system 66
Drought Prone Area Programme(DPAP) 32

Field Irrigation Requirement (FIR) 45
Flow irrigation system 14
Food and Agricultural Organisation (FAO) 2
Full Supply Depth (FSD) 102

Gross Irrigation Requirement (GIR)
Gross Command Area (GCA) 48

Integrated Soil and Water Improvement Project (ISAWIP) 176
International Programme for Technology and Research in Irrigation and Drainage (IPTRID) 259
Integrating the Multiple uses of Water (IWRM) 86
Integrated Water Resources Management (IWRM) 74
Irrigation water management 10

Lift irrigation system 15
Long-Crested Weir (LCW) 97

National Bank for Agriculture and Rural Development (NABARD)32
Net irrigation Requirement (NIR) 45

Parallel furrow drainage system 206
Participatory Irrigation Management (PIM) 9